SCRUM AND THE AGILE PLANNING ONION

SCUM STRATEGIC PLANNING
PORTFOLIO PLANNING
PRODUCT PLANNING
RELEASE PLANNING
ITERATION PLANNING

ANGELA SIRBU, MBA. PMP

© 2024 by Angela Sirbu, MBA. PMP. All rights reserved.

No part of this book may be reproduced or utilized in any form or by any means, electronic or mechanical, including photocopying, recording, or by any information storage and retrieval system, without permission in writing from the publisher.

First Edition 2024

Published by Angela Sirbu, MBA. PMP

CONTENTS

INTRODUCTION

CHAPTER 1: INTRODUCTION TO AGILE AND SCRUM

CHAPTER 2: STRATEGIC PLANNING IN SCRUM

CHAPTER 3: PORTFOLIO PLANNING

CHAPTER 4: PRODUCT PLANNING

CHAPTER 5: RELEASE PLANNING

CHAPTER 6: ITERATION PLANNING

CHAPTER 7: THE ROLE OF SCRUM MASTER

CHAPTER 8: THE ROLE OF PRODUCT OWNER

CHAPTER 9: THE DEVELOPMENT TEAM

CHAPTER 10: TOOLS AND TECHNIQUES

CHAPTER 11: SCALING AGILE AND SCRUM

CHAPTER 12: COMMON PITFALLS AND HOW TO AVOID THEM

CHAPTER 13: FUTURE TRENDS IN AGILE AND SCRUM

INTRODUCTION

In the dynamic world of software development, the ability to adapt and innovate is paramount. Agile methodologies, with Scrum at the forefront, offer a structured yet flexible framework to navigate the complexities of project management. This book, "Scrum and The Agile Planning Onion", delves into the intricate layers of planning that are essential for successful Agile implementation. From the broad strokes of strategic planning to the fine details of iteration planning, each layer plays a crucial role in ensuring that projects are not only completed on time but also deliver maximum value to stakeholders.

Understanding the various layers of planning within the Agile framework is akin to peeling an onion; each layer reveals new insights and requires different considerations. The outermost layer, strategic planning, aligns the project's goals with the organization's broader objectives. This high-level planning sets the stage for subsequent layers, ensuring that every decision made down the line supports the overarching vision. Moving inward, portfolio planning focuses on prioritizing and managing a collection of projects, balancing resources and timelines to achieve optimal outcomes.

Product planning, the next layer, zooms in on individual products, defining features and functionalities that meet user needs and market demands. This layer is critical for translating strategic goals into tangible deliverables. Release planning then breaks down these deliverables into manageable increments, allowing teams to deliver value early and often. This iterative approach not only enhances flexibility but also enables continuous feedback and improvement.

Finally, iteration planning, the innermost layer, involves detailed planning of individual sprints or iterations. This is where the rubber meets the road, as teams define specific tasks, assign responsibilities, and set short-term goals. Effective iteration planning ensures that each sprint is productive and aligned with the project's overall objectives.

"Scrum and The Agile Planning Onion" provides a comprehensive guide to mastering these layers of planning. It offers practical insights, real-world examples, and actionable strategies to help teams at every stage of the Agile journey. Whether you are a seasoned Scrum Master, a product owner, or a newcomer to Agile, this book will equip you with the knowledge and tools needed to navigate the complexities of Agile planning and deliver exceptional results.

Chapter 1: Introduction to Agile and Scrum

Agile Principles

Agile methodologies have gained prominence in contemporary project management, particularly within software development. The foundation of Agile is built on a set of principles that prioritize flexibility, collaboration, and customer satisfaction. These principles are designed to address the dynamic nature of project requirements and the need for frequent reassessment of outcomes. The Agile Manifesto, formulated in 2001, encapsulates these principles and serves as a guiding framework for Agile practices, including Scrum.

One of the core tenets of Agile principles is the prioritization of individuals and interactions over processes and tools. This principle underscores the significance of human elements in project management. Effective communication and collaboration among team members, stakeholders, and customers are deemed essential for the successful execution of projects. By fostering an environment where individuals can freely exchange ideas and feedback, Agile methodologies enhance the adaptive capacity of teams to respond to changes and challenges.

Working software is valued over comprehensive documentation in Agile principles. The emphasis on delivering functional software iteratively ensures that the product is continuously evolving and improving. This approach enables teams to provide tangible value to customers early and frequently. It also facilitates the identification and rectification of issues at an early stage, thereby reducing the risk of project failure. While documentation remains important, it is streamlined to support the primary goal of software delivery.

Customer collaboration is prioritized over contract negotiation. Agile principles advocate for a close partnership with customers throughout the project lifecycle. This continuous engagement ensures that the product aligns with customer needs and expectations. By involving customers in the development process, teams can gather valuable insights and make informed decisions. This collaborative approach mitigates the risk of misalignment between the delivered product and customer requirements.

Responding to change is preferred over following a plan. Agile principles recognize that project requirements are often subject to change due to various factors such as market dynamics, technological advancements, and evolving customer needs. Agile methodologies promote an iterative and incremental approach, allowing teams to adapt to changes swiftly. Regular reassessment of priorities and outcomes ensures that the project remains aligned with its goals and objectives.

The Agile principles also emphasize the importance of delivering value to customers through early and continuous delivery of valuable software. This principle is operationalized through short development cycles known as iterations or sprints. By delivering increments of the product at regular intervals, teams can gather feedback and make necessary adjustments. This iterative process enhances the quality and relevance of the final product.

Additionally, Agile principles advocate for sustainable development. Teams are encouraged to maintain a consistent pace of work, avoiding burnout and ensuring long-term productivity. This sustainable approach is achieved by balancing workload, setting realistic goals, and fostering a supportive work environment.

Continuous attention to technical excellence and good design is another principle of Agile. High-quality code, robust architecture, and effective design practices are essential for maintaining the integrity and scalability of the software. Agile methodologies promote practices such as refactoring, code reviews, and automated testing to uphold technical excellence.

Lastly, simplicity—the art of maximizing the amount of work not done—is a key principle in Agile. By focusing on essential features and eliminating unnecessary work, teams can streamline development processes and deliver more value with less effort. This principle encourages a minimalist approach to feature development, ensuring that the product remains lean and efficient.

These principles collectively form the foundation of Agile methodologies, guiding teams in delivering high-quality products in a dynamic and collaborative manner.

Overview of Scrum

Scrum, a subset of Agile methodology, constitutes a framework that facilitates iterative and incremental development of complex software projects. It is designed to deliver functional software in a time-boxed manner, through a series of sprints, each typically lasting two to four weeks. This approach is predicated on the principles of transparency, inspection, and adaptation, which collectively ensure continuous improvement and high-quality output.

The Scrum framework is composed of specific roles, events, and artifacts that work in tandem to streamline the development process. The primary roles within Scrum include the Product Owner, the Scrum Master, and the Development Team. The Product Owner is responsible for defining the features of the product and prioritizing the backlog to maximize value. The Scrum Master acts as a facilitator, ensuring that the team adheres to Scrum practices and removing any impediments that may hinder progress. The Development Team, a cross-functional group of professionals, is tasked with delivering potentially shippable increments of the product at the end of each sprint.

Central to Scrum are its events, which provide structure and regular opportunities for the team to inspect and adapt. These events include the Sprint Planning Meeting, the Daily Scrum, the Sprint Review, and the Sprint Retrospective. The Sprint Planning Meeting marks the beginning of a sprint, during which the team collaborates to select items from the product backlog and commits to completing them. The Daily Scrum, a short, time-boxed meeting held each day, allows the team to synchronize activities and identify any obstacles. The Sprint Review, conducted at the end of the sprint, involves the team presenting the completed work to stakeholders for feedback. The Sprint Retrospective follows the Sprint Review and provides the team with an opportunity to reflect on the sprint and identify areas for improvement.

Scrum artifacts serve to ensure transparency and provide a clear understanding of the work being undertaken. These include the Product Backlog, the Sprint Backlog, and the Increment. The Product Backlog is an ordered list of all desired work on the project, maintained by the Product Owner. The Sprint Backlog consists of the items selected for the current sprint, along with a plan for delivering them. The Increment is the sum of all completed product backlog items at the end of a sprint, representing the latest version of the product.

Scrum's empirical process control model is rooted in the idea that knowledge comes from experience and decision-making based on what is known. This model relies on frequent inspection and adaptation to cope with complex and unpredictable environments. By promoting transparency, Scrum ensures that all aspects of the process are visible to those responsible for the outcome. Regular inspection allows the team to assess progress and detect variances early. Adaptation enables the team to make necessary adjustments to optimize performance and achieve the desired results.

The efficacy of Scrum is supported by numerous studies and real-world applications, demonstrating its ability to enhance productivity, improve product quality, and increase stakeholder satisfaction. Its structured yet flexible approach allows teams to respond to changing requirements and deliver valuable software incrementally. Consequently, Scrum has become a widely adopted framework in the software development industry, offering a robust mechanism for managing complex projects in an agile manner.

Benefits of Agile Planning

The Agile Planning Onion, a conceptual framework for iterative and incremental planning, offers numerous advantages that enhance both the efficiency and effectiveness of project management. The first notable benefit is the increased adaptability to changing requirements. Traditional planning methodologies often struggle with scope changes, leading to project delays and budget overruns. Agile planning, with its iterative cycles and feedback loops, allows teams to incorporate new information and pivot as necessary. This dynamic adaptability ensures that the project remains aligned with stakeholder needs and market conditions.

Another significant advantage is improved stakeholder engagement and satisfaction. Agile planning methodologies, such as Scrum, prioritize frequent communication and collaboration with stakeholders. Regular reviews and demonstrations provide stakeholders with visibility into the project's progress and an opportunity to offer feedback. This continuous engagement fosters a sense of ownership and ensures that the final product meets or exceeds expectations. Additionally, the transparency inherent in Agile planning builds trust between the development team and stakeholders, which is crucial for long-term project success.

Enhanced team collaboration and morale are also key benefits of Agile planning. The emphasis on self-organizing teams and collective decision-making empowers team members, fostering a culture of accountability and mutual respect. Daily stand-up meetings, sprint planning sessions, and retrospectives facilitate open communication and ensure that any issues are promptly addressed. This collaborative environment not only enhances productivity but also contributes to higher job satisfaction and reduced turnover rates among team members.

The iterative nature of Agile planning results in early and continuous delivery of valuable increments. By breaking down the project into smaller, manageable units of work, teams can deliver functional components at the end of each sprint. This incremental delivery approach allows for early detection of defects, reducing the risk of significant issues arising late in the project lifecycle. Furthermore, the early delivery of usable increments enables stakeholders to realize value sooner, which can be a critical competitive advantage in fast-paced markets.

Risk management is another area where Agile planning excels. The frequent reassessment of project priorities and risks, coupled with the iterative delivery of increments, allows teams to identify and mitigate potential issues early. This proactive approach to risk management reduces the likelihood of major setbacks and ensures that the project remains on track. Additionally, the practice of conducting retrospectives at the end of each sprint provides an opportunity for continuous improvement, enabling teams to learn from past experiences and refine their processes.

The Agile Planning Onion also promotes a focus on delivering customer value. By prioritizing work based on stakeholder input and business value, Agile teams ensure that the most critical and valuable features are developed first. This value-driven approach maximizes return on investment and ensures that resources are allocated efficiently. Moreover, the flexibility to reprioritize work based on changing needs ensures that the project remains relevant and aligned with business objectives.

In terms of scalability, Agile planning frameworks like the Agile Planning Onion can be adapted to suit projects of varying sizes and complexities. Whether applied to small teams or large, multi-team initiatives, the principles of iterative planning, continuous feedback, and stakeholder collaboration remain effective. This scalability makes Agile planning a versatile tool for organizations across different industries and project types.

The cumulative effect of these benefits is a more responsive, efficient, and effective project management approach. By fostering adaptability, enhancing stakeholder engagement, improving team collaboration, delivering value incrementally, managing risks proactively, and focusing on customer value, Agile planning offers a robust framework for successful project execution.

The Agile Planning Onion

Agile methodologies have gained prominence as effective frameworks for managing complex projects, with Scrum being one of the most widely adopted. A critical aspect of Scrum and Agile processes is planning, which is often misunderstood as a single, monolithic task. In reality, Agile planning is a multi-layered activity, conceptualized effectively through the metaphor of the Agile Planning Onion. This model delineates planning into several distinct layers, each serving a unique purpose and time horizon, thereby facilitating a more structured approach to project management.

The outermost layer of the Agile Planning Onion is the **Product Vision**. This layer serves as the overarching goal or the long-term objective of the project. It is typically broad and strategic, providing a high-level understanding of what the project aims to achieve. The Product Vision is essential for aligning all stakeholders and ensuring that the project consistently progresses towards a shared goal. This vision is usually articulated by the Product Owner and requires minimal revisions once established, serving as a guiding star throughout the project's lifecycle.

Moving inward, the next layer is the **Product Roadmap**. This layer translates the Product Vision into a more tangible and time-bound plan. The Product Roadmap outlines major milestones, significant deliverables, and key timelines. It provides a temporal framework that helps in tracking progress and making strategic decisions. The roadmap is dynamic and may evolve based on market conditions, stakeholder feedback, or technological advancements. However, it remains less granular than the subsequent planning layers, maintaining a focus on medium to long-term objectives.

The **Release Planning** layer is situated further inward. This layer breaks down the Product Roadmap into specific releases, each of which delivers a subset of the product's functionality. Release Planning is crucial for delivering incremental value to stakeholders and users. It involves prioritizing features, setting deadlines, and allocating resources for each release. This layer requires regular updates and revisions to adapt to changing requirements and to incorporate feedback from previous releases.

The penultimate layer is **Iteration Planning**. This layer is concerned with the short-term, typically spanning one to four weeks, depending on the iteration length defined by the Scrum team. Iteration Planning translates the deliverables of the Release Plan into actionable tasks for the team. It involves detailed planning of the work to be accomplished during the iteration, with a focus on achieving specific, measurable goals. This layer is highly dynamic, requiring frequent adjustments to address impediments, incorporate new information, and ensure continuous delivery of value.

At the core of the Agile Planning Onion is the **Daily Planning** layer. This innermost layer is characterized by daily stand-up meetings, where team members synchronize their activities, discuss progress, and identify any impediments. Daily Planning ensures that the team remains aligned and can swiftly adapt to emerging challenges. It fosters a culture of continuous communication and collaboration, which is essential for maintaining momentum and ensuring the successful execution of the iteration plan.

Each layer of the Agile Planning Onion is interconnected, with the outer layers providing context and direction for the inner layers. This multi-layered approach enables Scrum teams to plan effectively across different time horizons, ensuring that strategic objectives are met while maintaining the flexibility to adapt to change. Understanding and implementing the Agile Planning Onion can significantly enhance the effectiveness of Agile project management, leading to more predictable and successful outcomes.

Chapter 2: Strategic Planning in Scrum

Defining Strategic Goals

Strategic goals are fundamental to the success of any organization utilizing Scrum and the Agile Planning Onion framework. These goals provide a clear direction and set the foundation for all subsequent planning and execution activities. The establishment of strategic goals involves a comprehensive understanding of the organization's vision, mission, and long-term objectives. This subchapter delineates the critical components and methodologies for effectively defining strategic goals within an Agile context.

The initial step in defining strategic goals requires a thorough analysis of the organization's vision and mission statements. The vision statement outlines the long-term aspirations and desired future state of the organization, while the mission statement articulates the core purpose and primary objectives. Together, these statements serve as guiding principles for the formulation of strategic goals. It is essential that these goals are aligned with the broader organizational vision to ensure coherence and consistency in strategic planning.

Subsequent to the articulation of vision and mission statements, a situational analysis is conducted to assess the internal and external environments of the organization. This analysis typically employs frameworks such as SWOT (Strengths, Weaknesses, Opportunities, Threats) or PESTLE (Political, Economic, Social, Technological, Legal, Environmental) to identify key factors that may impact the achievement of strategic goals. The insights derived from this analysis inform the prioritization and refinement of strategic objectives.

Once the situational analysis is complete, the next phase involves setting SMART (Specific, Measurable, Achievable, Relevant, Time-bound) goals. Specificity ensures that goals are clear and unambiguous, facilitating focused effort and resource allocation. Measurability allows for the tracking of progress and performance, enabling data-driven decision-making. Achievability ensures that goals are realistic and attainable within the given constraints. Relevance ensures that goals are aligned with the overall strategic direction of the organization. Time-bound goals provide a clear timeframe for achievement, fostering a sense of urgency and accountability.

In the context of Scrum and the Agile Planning Onion, strategic goals must also be adaptable to changing circumstances. The Agile methodology emphasizes flexibility and responsiveness, necessitating that strategic goals are periodically reviewed and adjusted as needed. This iterative approach ensures that the organization remains aligned with its strategic objectives while adapting to evolving market conditions and internal dynamics.

Furthermore, the involvement of key stakeholders in the goal-setting process is crucial. Engaging stakeholders, including executives, team members, and customers, ensures that diverse perspectives are considered and that the strategic goals reflect the collective aspirations of the organization. Collaborative goal-setting fosters a sense of ownership and commitment among stakeholders, enhancing the likelihood of successful implementation.

To operationalize strategic goals, they must be cascaded down through the organization, translating high-level objectives into actionable plans at various levels. This cascading process involves breaking down strategic goals into tactical and operational goals, aligning them with specific teams and individuals. The Agile Planning Onion framework facilitates this alignment by providing a structured approach to planning at different levels of granularity, from broad strategic goals to detailed task-level plans.

In conclusion, defining strategic goals within the Scrum and Agile Planning Onion framework is a multifaceted process that requires a clear understanding of the organization's vision and mission, a thorough situational analysis, the formulation of SMART goals, adaptability, stakeholder engagement, and effective cascading of goals. By adhering to these principles, organizations can establish a solid foundation for successful Agile planning and execution.

Aligning with Organizational Vision

In the context of Scrum and Agile methodologies, aligning team activities with the organizational vision is a critical factor for achieving cohesive and effective project outcomes. The organizational vision provides a strategic framework that guides decision-making processes, prioritization, and resource allocation. A clear alignment ensures that all efforts contribute to the overarching goals of the organization, thereby enhancing both efficiency and impact.

The organizational vision represents the long-term objectives and aspirations of a company. It serves as a beacon for all stakeholders, including teams working within the Scrum framework. When Scrum teams understand and internalize this vision, they can better align their work processes and deliverables with strategic priorities. This alignment is achieved through several mechanisms embedded within the Scrum and Agile planning processes.

One of the primary mechanisms for fostering alignment is the Product Backlog. The Product Owner plays a pivotal role in translating the organizational vision into actionable items within the Product Backlog. By prioritizing backlog items that are most aligned with the strategic goals, the Product Owner ensures that the team's efforts are directed towards high-impact areas. This prioritization is not a one-time activity but an ongoing process that requires constant communication with stakeholders and regular refinement sessions.

Another critical aspect is the Sprint Planning meeting, where the Scrum team selects items from the Product Backlog to work on during the upcoming sprint. During this meeting, the Product Owner communicates the rationale behind the prioritization of tasks, linking them explicitly to the organizational vision. This context-setting exercise helps the team understand the strategic importance of their work, fostering a sense of purpose and motivation.

Furthermore, the Daily Scrum meetings provide an opportunity for continuous alignment. These short, focused meetings allow team members to synchronize their activities and identify any impediments that may hinder progress towards the sprint goals. By keeping the organizational vision in mind, team members can make informed decisions about task prioritization and resource allocation on a day-to-day basis.

The Sprint Review and Sprint Retrospective meetings also play a significant role in maintaining alignment with the organizational vision. During the Sprint Review, the team demonstrates the completed work to stakeholders, receiving feedback that can be used to adjust future priorities and ensure ongoing alignment. The Sprint Retrospective, on the other hand, allows the team to reflect on their processes and outcomes, identifying areas for improvement that can enhance their ability to deliver value in line with the organizational vision.

Effective communication is essential for maintaining alignment. Regular updates from the leadership team about strategic shifts or evolving priorities help Scrum teams stay informed and agile. Additionally, fostering a culture of transparency and open dialogue ensures that all team members feel connected to the organizational vision and are empowered to contribute towards it.

In conclusion, aligning Scrum activities with the organizational vision is a dynamic and iterative process that requires active involvement from all team members and stakeholders. By embedding the vision into the core Scrum ceremonies and fostering a culture of continuous communication and reflection, organizations can ensure that their Agile initiatives drive meaningful and strategic outcomes. This alignment not only enhances the effectiveness of Scrum teams but also contributes to the overall success and competitiveness of the organization.

Long-term Planning

The process of long-term planning within the context of Scrum and the Agile Planning Onion requires a strategic and systematic approach, integrating both flexibility and foresight. Unlike traditional methodologies that often rely on rigid, predictive models, Agile frameworks necessitate a more adaptable and iterative approach to accommodate evolving project requirements and market conditions.

A crucial element of long-term planning in Agile is the establishment of a coherent vision. This vision serves as a guiding beacon, aligning the project's trajectory with the overarching organizational goals. It is imperative to engage key stakeholders early in the planning phase to ensure that their insights and expectations are incorporated. This collaborative effort helps in creating a shared understanding and commitment to the project's objectives.

Strategic themes or pillars are identified to break down the vision into manageable components. These themes represent high-level goals or areas of focus that need to be addressed to realize the vision. Each theme is further decomposed into initiatives or epics, which are substantial bodies of work that can span multiple sprints or even releases. Initiatives are prioritized based on their value contribution, feasibility, and alignment with the strategic themes.

A roadmap is then developed to outline the sequence and timing of these initiatives. This roadmap is not a fixed plan but a living document that evolves based on empirical data and feedback. It provides a temporal perspective, illustrating how the various initiatives will unfold over time and how they interconnect to drive the project towards its vision. The roadmap helps in setting expectations and provides a framework for tracking progress and making adjustments as necessary.

Capacity planning is another critical aspect of long-term planning. It involves assessing the team's ability to deliver on the planned initiatives within the given timeframes. This assessment takes into account the team's velocity, resource availability, and potential constraints. By understanding capacity, teams can make informed decisions about what is feasible and adjust the scope or timelines accordingly. This proactive approach helps in mitigating risks and avoiding overcommitment.

Risk management is inherently integrated into the long-term planning process. Identifying potential risks and developing mitigation strategies ensures that the project remains resilient in the face of uncertainties. Regular risk reviews and adjustments to the plan based on emerging risks are essential to maintain alignment with the project's goals.

Continuous feedback loops play a pivotal role in refining long-term plans. Regular reviews, retrospectives, and stakeholder feedback sessions provide valuable insights that inform adjustments to the roadmap and strategic themes. This iterative refinement ensures that the plan remains relevant and responsive to changing circumstances.

The integration of metrics and key performance indicators (KPIs) is essential for monitoring progress and assessing the effectiveness of the long-term plan. These metrics provide quantitative data that can be analyzed to identify trends, measure performance, and make data-driven decisions. Common metrics include delivery rate, quality indicators, and customer satisfaction scores.

Long-term planning in Agile is a dynamic and collaborative process that balances strategic foresight with the flexibility to adapt to change. By maintaining a clear vision, breaking down work into manageable components, developing a responsive roadmap, assessing capacity, managing risks, and leveraging continuous feedback, organizations can effectively navigate the complexities of large-scale projects and drive sustained success.

Case Studies in Strategic Planning

The integration of Scrum within the Agile Planning Onion framework has led to significant advancements in strategic planning across various industries. This subchapter examines three distinct case studies to highlight the practical application and benefits of this integration.

The first case study involves a large-scale software development company seeking to enhance its product delivery timelines. Initially, the company struggled with prolonged development cycles, leading to delayed product launches and customer dissatisfaction. By adopting Scrum within the Agile Planning Onion, the company restructured its planning process into five layers: product vision, product roadmap, release planning, iteration planning, and daily planning. This stratified approach facilitated clearer communication of strategic goals across all levels of the organization. The iterative nature of Scrum, combined with the layered planning onion, allowed for continuous feedback and adjustments, significantly reducing time-to-market and improving product quality.

In the second case study, a healthcare provider aimed to streamline its patient management system. The existing system was fraught with inefficiencies and lacked coordination among different departments. Implementing Scrum practices within the Agile Planning Onion framework enabled the healthcare provider to break down the project into manageable sprints. The product vision was aligned with the overall goal of improving patient care, while the product roadmap detailed specific features required to achieve this vision. Release planning ensured that each phase of the project delivered incremental value, while iteration planning allowed for regular adjustments based on stakeholder feedback. Daily planning meetings facilitated real-time problem-solving and enhanced team collaboration. The result was a more efficient patient management system that improved service delivery and patient satisfaction.

The third case study examines a financial services firm that sought to modernize its legacy systems. The firm faced challenges with outdated technology, which hindered its ability to offer competitive financial products. By leveraging Scrum within the Agile Planning Onion, the firm established a clear product vision focused on digital transformation. The product roadmap outlined key milestones for system upgrades and new feature implementations. Release planning ensured that each phase of the modernization effort delivered tangible improvements, while iteration planning allowed the team to adapt to changing requirements and technological advancements. Daily planning sessions fostered a collaborative environment where team members could address issues promptly. The strategic planning approach resulted in a successful modernization of the firm's systems, enabling it to offer innovative financial products and services.

These case studies illustrate the efficacy of integrating Scrum within the Agile Planning Onion for strategic planning. The structured yet flexible nature of this approach allows organizations to align their strategic goals with actionable plans, fostering continuous improvement and innovation. By breaking down complex projects into manageable layers and incorporating iterative feedback loops, organizations can navigate uncertainties and deliver value incrementally. This methodology not only enhances project outcomes but also promotes a culture of collaboration and adaptability, essential for thriving in today's dynamic business environment.

The practical application of Scrum within the Agile Planning Onion framework demonstrates its potential to transform strategic planning across diverse sectors. Whether in software development, healthcare, or financial services, this integrated approach offers a robust solution for addressing complex challenges and achieving strategic objectives.

Chapter 3: Portfolio Planning

Understanding Portfolio Management

Portfolio management within the realm of Scrum and Agile methodologies necessitates a nuanced understanding of how to effectively align projects with overarching organizational goals. The Agile Planning Onion offers a structured approach to this alignment, where portfolio management serves as a critical layer. This layer ensures strategic coherence and resource optimization across multiple projects.

In the context of Agile, portfolio management transcends traditional project oversight by emphasizing adaptability, continuous value delivery, and iterative improvement. The primary objective is to create a dynamic portfolio that can respond swiftly to market changes and internal shifts. This requires a robust framework for prioritization, resource allocation, and performance monitoring, all while maintaining alignment with strategic objectives.

One of the foundational principles of Agile portfolio management is the concept of value streams. Value streams represent the sequence of activities that deliver value to the customer. By focusing on these streams, organizations can better identify and prioritize initiatives that contribute the most to their strategic goals. This focus helps in minimizing waste and ensuring that resources are directed towards the most impactful projects.

Effective portfolio management in an Agile environment also involves the continuous reprioritization of projects based on current data and feedback. This iterative process is facilitated through regular review cycles, often synchronized with the sprint or iteration cycles of individual projects. The goal is to maintain a flexible and responsive portfolio that can adapt to new information and changing circumstances.

Another critical aspect is the alignment of portfolio management with organizational strategy. This alignment is achieved through strategic themes or objectives that guide the selection and prioritization of initiatives. These strategic themes are derived from the organization's long-term goals and are used to evaluate the potential impact of each project. By ensuring that every project within the portfolio supports these themes, organizations can create a cohesive strategy that drives overall business success.

Resource allocation is a key challenge in portfolio management. In Agile, this involves not just the distribution of financial resources but also the allocation of human resources and time.

Agile portfolio management requires a dynamic approach to resource allocation, where resources are reallocated based on project needs and priorities. This flexibility allows organizations to optimize their resource usage and ensure that critical projects receive the necessary support.

Performance monitoring and metrics are essential for effective portfolio management. Agile methodologies advocate for the use of key performance indicators (KPIs) and other metrics to track the progress and success of projects within the portfolio. These metrics provide valuable insights into project performance, helping managers make informed decisions about which projects to continue, modify, or terminate.

The role of leadership in Agile portfolio management cannot be overstated. Leaders must foster a culture of transparency, collaboration, and continuous improvement. They need to facilitate open communication channels and ensure that teams have the autonomy to make decisions and adapt to changes. By empowering teams and promoting a culture of agility, leaders can enhance the effectiveness of portfolio management.

Incorporating Agile principles into portfolio management also involves embracing a mindset of experimentation and learning. Organizations should be willing to try new approaches, learn from failures, and continuously refine their strategies. This iterative learning process is fundamental to Agile and is crucial for maintaining a responsive and adaptive portfolio.

In summary, portfolio management within the Agile Planning Onion framework requires a strategic, flexible, and data-driven approach. By focusing on value streams, aligning with strategic objectives, dynamically allocating resources, and continuously monitoring performance, organizations can effectively manage their portfolios in a way that drives sustained value and supports their long-term goals.

Prioritization Techniques

Effective prioritization is a cornerstone of successful Scrum and Agile methodologies, directly influencing the efficiency and outcome of project delivery. Various techniques have been developed to assist teams in making informed decisions regarding the sequencing of tasks, features, and user stories. These techniques aim to optimize value delivery, mitigate risks, and ensure alignment with business objectives.

One widely recognized method is the MoSCoW prioritization framework, which categorizes items into four distinct groups: Must have, Should have, Could have, and Won't have. This approach facilitates clear communication among stakeholders by delineating essential requirements from those that are desirable but not critical. The "Must have" category includes non-negotiable elements essential for project success, whereas "Should have" items are important but not imperative for the initial release. "Could have" features enhance the product but are not crucial, and "Won't have" items are agreed upon as not being included in the current iteration. This stratification aids in managing scope and expectations effectively.

Another technique, the Kano Model, evaluates features based on customer satisfaction and their impact on user experience. Features are classified into five categories: Basic needs, Performance needs, Excitement needs, Indifferent attributes, and Reverse attributes. Basic needs are fundamental and their absence leads to dissatisfaction, while Performance needs correlate directly with customer satisfaction. Excitement needs, although not expected, can significantly enhance user satisfaction when present. Indifferent attributes have little impact on satisfaction, and Reverse attributes can cause dissatisfaction if included. The Kano Model helps teams focus on features that maximize customer satisfaction and differentiate the product in the market.

The Value vs. Effort matrix is another pragmatic tool used in Agile prioritization. By plotting tasks or features on a two-dimensional grid based on their perceived value and the effort required for implementation, teams can identify high-value, low-effort items that should be prioritized. This visual representation aids in strategic decision-making, ensuring that resources are allocated efficiently to maximize return on investment.

Cost of Delay (CoD) and Weighted Shortest Job First (WSJF) are techniques grounded in economic principles. CoD quantifies the economic impact of delaying a task or feature, emphasizing the urgency of high-impact items. WSJF, on the other hand, calculates a priority score by dividing the CoD by the job size (effort). This ratio helps prioritize tasks that deliver the highest economic value in the shortest time, promoting a balance between value delivery and resource utilization.

User Story Mapping is a collaborative approach that visualizes the user journey and aligns features with user needs. This technique involves creating a map of user activities and breaking them down into smaller tasks or stories. By organizing these stories into a sequence that reflects the user experience, teams can prioritize features that deliver the most significant value to the end-user, ensuring a user-centric development process.

Incorporating these prioritization techniques within the framework of the Agile Planning Onion enhances the ability to make informed, strategic decisions. The Agile Planning Onion, which encompasses layers of planning from daily activities to long-term vision, provides a structured approach to planning at different levels of granularity. By integrating prioritization techniques at each layer, teams can ensure that both short-term iterations and long-term goals are aligned with the overarching business objectives, fostering a cohesive and adaptive planning process.

Resource Allocation

Effective resource allocation is paramount in Scrum and the Agile planning onion framework. This process involves systematically distributing available resources—such as time, personnel, and budget—towards tasks and projects to maximize efficiency and achieve strategic objectives. In Agile methodologies, the fluid and iterative nature of project management necessitates a dynamic approach to resource allocation, ensuring that resources are continuously realigned to meet evolving project requirements and stakeholder expectations.

Scrum teams operate within time-boxed iterations known as sprints. Each sprint begins with a planning meeting where the team commits to deliverables based on their capacity and the priority of tasks. The capacity is often determined by historical data, such as velocity, which measures the amount of work a team can complete in a given sprint. This historical data aids in forecasting future performance and in making informed decisions about resource allocation. Key resources, including team members with specific skill sets, are assigned to tasks that match their expertise, ensuring that the most critical and complex tasks receive the attention they require.

The Agile planning onion provides a multi-layered approach to planning, with each layer representing a different level of granularity. Starting from the outermost layer, which focuses on the product vision, down to the innermost layer, which deals with daily tasks, resource allocation must be managed at each level to maintain alignment with overall goals. At the product vision level, resources are allocated based on long-term strategic objectives, ensuring that the project's direction aligns with the organization's mission and market demands. As planning progresses inward, resources are distributed with increasing specificity, focusing on detailed tasks and immediate deliverables.

One of the critical challenges in resource allocation within Scrum is balancing the competing demands of multiple projects and priorities. This is typically managed through backlog refinement sessions, where the product backlog is continuously reviewed and prioritized. High-priority items are allocated resources first, ensuring that the most valuable features and tasks are addressed promptly. This prioritization is often guided by the product owner, who acts as the liaison between stakeholders and the development team, ensuring that the allocation of resources reflects the strategic importance of each task.

Resource allocation also involves managing constraints and dependencies. In many cases, tasks are interdependent, requiring careful coordination to ensure that the completion of one task does not delay others. Scrum Masters play a pivotal role in this process by facilitating communication and collaboration among team members, identifying potential bottlenecks, and ensuring that resources are reallocated as needed to maintain workflow efficiency.

Furthermore, Agile methodologies emphasize the importance of flexibility and adaptability in resource allocation. As project requirements and external conditions change, so too must the distribution of resources. This adaptability is achieved through regular sprint reviews and retrospectives, where the team assesses their performance and identifies areas for improvement. Adjustments to resource allocation are made based on these insights, allowing the team to respond proactively to challenges and changes.

In essence, effective resource allocation in Scrum and the Agile planning onion requires a balanced approach that considers both strategic objectives and immediate operational needs. By leveraging historical data, prioritizing tasks, managing dependencies, and maintaining flexibility, Scrum teams can optimize their resource utilization and enhance their ability to deliver high-quality products efficiently.

Measuring Success

Effective measurement is a cornerstone of the Scrum methodology and is critical for evaluating the success of Agile planning. The Agile Planning Onion provides a structured approach to planning, but without robust metrics, it becomes challenging to gauge progress and identify areas for improvement. This subchapter focuses on the various metrics and key performance indicators (KPIs) that are essential for assessing the effectiveness of Scrum practices within the Agile Planning Onion framework.

Scrum teams primarily rely on a set of quantitative and qualitative metrics to measure success. These metrics can be broadly categorized into four types: delivery, performance, quality, and team health. Delivery metrics include velocity, sprint burndown, and release burndown charts. Velocity measures the amount of work a team can complete in a given sprint, providing insight into the team's capacity and helping to predict future performance. Sprint burndown charts track the progress of work within a sprint, highlighting any deviations from the planned trajectory. Release burndown charts extend this concept to the entire release cycle, offering a macro-level view of progress towards project completion.

Performance metrics focus on the efficiency and effectiveness of the Scrum process. These include cycle time, lead time, and throughput. Cycle time measures the duration from the start to the end of a task, while lead time encompasses the entire timeline from task creation to task completion. Throughput quantifies the number of tasks completed in a given timeframe. Analyzing these metrics helps identify bottlenecks and inefficiencies, enabling teams to optimize their workflows.

Quality metrics are pivotal for ensuring that the end product meets the desired standards. Defect density, code coverage, and customer satisfaction scores are commonly used in this category. Defect density measures the number of defects per unit of code, providing insights into the code quality. Code coverage assesses the extent to which the codebase is tested, ensuring that critical paths are adequately covered by automated tests. Customer satisfaction scores, often gathered through surveys and feedback loops, offer direct insights into the user's perspective on the product's quality and usability.

Team health metrics are essential for sustaining long-term productivity and morale. These include team satisfaction, collaboration effectiveness, and burnout rates. Team satisfaction surveys gauge the overall happiness and engagement of team members, while collaboration effectiveness can be measured through tools like the Team Collaboration Index. Burnout rates, often assessed through regular check-ins and health surveys, help identify teams at risk of burnout, allowing for timely interventions.

In addition to these metrics, the concept of empirical process control underpins the Scrum framework. This involves regular inspection and adaptation based on the collected data. Sprint retrospectives, for instance, are crucial ceremonies where teams reflect on their performance and identify actionable improvements based on the metrics gathered. These retrospectives foster a culture of continuous improvement, aligning with the Agile principle of iterative development.

However, it is important to acknowledge that metrics should not be viewed in isolation. A holistic approach, considering multiple metrics in conjunction, provides a more accurate picture of success. For instance, a high velocity might seem positive, but if it coincides with a high defect density, it indicates a trade-off between speed and quality. Therefore, balancing these metrics is key to achieving sustainable success in Scrum practices.

The Agile Planning Onion framework, with its layered approach to planning, benefits significantly from these metrics. By providing clear, actionable insights at each layer—from strategic planning to daily activities—metrics ensure that teams remain aligned with their goals and can adapt swiftly to changing circumstances. Through continuous measurement and improvement, Scrum teams can enhance their efficiency, product quality, and overall satisfaction, ultimately leading to successful project outcomes.

Chapter 4: Product Planning

Product Vision and Roadmap

The foundation of successful product development in Agile methodologies, particularly within the Scrum framework, is a well-defined product vision and a meticulously crafted roadmap. A product vision serves as a strategic guide, encapsulating the essence of what the product aims to achieve and the value it intends to deliver to its users. This vision must be clear, compelling, and aligned with the broader organizational goals. It acts as a North Star, ensuring that all stakeholders, including the development team, product owner, and customers, are aligned in their understanding of the product's purpose and direction.

The product vision is not merely a statement but a profound declaration of intent. It typically encapsulates the target market, the problem the product aims to solve, and the key differentiators that set it apart from competitors. Crafting an effective product vision requires a deep understanding of the market landscape, user needs, and competitive dynamics. It often involves extensive research, user interviews, and iterative refinement to ensure that it resonates with the intended audience and stakeholders.

Once the product vision is established, the next critical step is to create a product roadmap. The roadmap translates the strategic vision into a tactical plan, outlining the high-level milestones and deliverables that will bring the vision to fruition. A well-constructed roadmap provides a temporal perspective, delineating the sequence of development activities and major releases. It serves as a communication tool, ensuring transparency and alignment among all stakeholders regarding the product's development trajectory.

In the context of Scrum, the product roadmap is inherently flexible, accommodating changes based on feedback and evolving market conditions. This adaptability is a core tenet of Agile methodologies, emphasizing the importance of responsiveness and iterative progress over rigid, long-term planning. The roadmap typically spans multiple sprints, providing a high-level overview of the product's evolution while allowing for detailed planning at the sprint level.

The product owner plays a pivotal role in both defining the product vision and maintaining the roadmap. This individual is responsible for ensuring that the vision is clearly articulated and communicated to the development team and other stakeholders. The product owner also prioritizes the product backlog based on the roadmap, ensuring that the most valuable features and enhancements are delivered first. This prioritization process is dynamic, continually reassessed based on user feedback, market changes, and technological advancements.

Effective collaboration between the product owner, Scrum Master, and development team is essential for the successful realization of the product vision and roadmap. The Scrum Master facilitates this collaboration, ensuring that the team adheres to Scrum principles and practices, and that impediments to progress are swiftly addressed. The development team, empowered by a clear vision and a coherent roadmap, can then focus on delivering high-quality increments that incrementally build towards the product's strategic objectives.

In summary, the product vision and roadmap are integral components of the Agile planning onion, providing the strategic and tactical foundations for product development within the Scrum framework. They ensure that all efforts are aligned with the overarching goals, enabling the team to deliver value consistently and effectively. Through continuous refinement and stakeholder engagement, the product vision and roadmap remain relevant and actionable, guiding the product's journey from concept to market success.

User Stories and Backlogs

User Stories and Backlogs are fundamental components within the Scrum framework, serving as the primary mechanisms for capturing and organizing requirements. User Stories are concise, simple descriptions of a feature or a functionality from the perspective of the end user. They are typically structured using the format: "As a [type of user], I want [an action] so that [a benefit/a value]." This format ensures that the focus remains on the user's needs and the value delivered to them, aligning with the core principles of Agile methodologies.

The process of creating User Stories involves collaboration between the Product Owner, stakeholders, and the development team. This collaborative effort ensures that the requirements are clearly understood and agreed upon by all parties. User Stories should be INVEST (Independent, Negotiable, Valuable, Estimable, Small, and Testable) to ensure they are effective in driving the development process. Independence ensures that stories can be developed in any order, negotiability allows for flexibility during implementation, value guarantees that each story delivers a tangible benefit, estimability ensures that the team can gauge the effort required, smallness facilitates rapid completion, and testability ensures that the acceptance criteria are clear and verifiable.

The Product Backlog is a prioritized list of all the work that needs to be done within the project. It is dynamic, constantly evolving as new information emerges and priorities shift. The Product Owner is responsible for maintaining and prioritizing the Product Backlog, ensuring that it reflects the current needs and goals of the project. Items in the Product Backlog can range from high-level epics to detailed User Stories, technical tasks, or bugs. Each item in the backlog is typically accompanied by a description, priority, and an estimate of the effort required.

Backlog refinement, also known as grooming, is a regular activity within Scrum where the Product Owner and the development team review and update the Product Backlog. This process involves breaking down high-level items into smaller, more manageable User Stories, re-estimating effort, and re-prioritizing based on the latest insights. Regular backlog refinement sessions help ensure that the backlog remains relevant and actionable, preventing it from becoming an unwieldy list of outdated or irrelevant items.

Sprint Backlogs are subsets of the Product Backlog, containing the User Stories and tasks that the development team commits to completing within a specific sprint. The Sprint Backlog is created during the Sprint Planning meeting, where the team selects items from the top of the Product Backlog based on their priority and the team's capacity. The Sprint Backlog is a living artifact, with the team updating it throughout the sprint to reflect progress and any changes that may occur.

Effective management of User Stories and Backlogs is critical for the success of Scrum projects. It ensures that the development efforts are aligned with the overall project goals and that the team remains focused on delivering high-value features to the end users. By maintaining clear, well-defined User Stories and a prioritized, actionable Product Backlog, Scrum teams can navigate the complexities of software development with agility and precision.

Stakeholder Involvement

Stakeholder engagement is a critical component in the successful implementation of Scrum within the Agile Planning Onion framework. The involvement of stakeholders ensures that the product development process remains aligned with the strategic objectives of the organization and the needs of the end-users. This alignment is achieved by fostering transparent communication, continuous feedback, and collaborative decision-making throughout the development lifecycle.

The role of stakeholders in Scrum is multifaceted, encompassing various responsibilities and interactions. Key stakeholders typically include customers, end-users, product owners, project managers, and executive sponsors. Each of these roles brings unique perspectives and requirements to the table, which must be integrated into the planning and execution phases.

One of the fundamental principles of Scrum is the inclusion of stakeholders in the iterative review process. During Sprint Reviews, stakeholders are invited to inspect the product increments and provide feedback. This regular inspection allows for the early detection of issues and the incorporation of changes that enhance the product's value. By actively participating in these reviews, stakeholders can ensure that the product remains relevant and meets their evolving needs.

The Product Owner plays a pivotal role in bridging the gap between the development team and the stakeholders. Acting as the liaison, the Product Owner is responsible for gathering requirements, prioritizing the product backlog, and ensuring that the development team understands the stakeholder's vision and goals. This role necessitates a deep understanding of both the business context and the technical aspects of the product. Effective communication skills are essential for the Product Owner to articulate the stakeholders' needs clearly and to manage their expectations.

In addition to formal meetings such as Sprint Reviews and Planning Sessions, informal interactions between stakeholders and the development team are equally important. These interactions can take the form of ad-hoc discussions, collaborative workshops, or even informal feedback sessions. Such engagements foster a culture of openness and continuous improvement, where ideas can be freely exchanged, and innovative solutions can emerge.

The Agile Planning Onion model emphasizes the importance of involving stakeholders at multiple levels of planning. At the strategic level, stakeholders contribute to defining the long-term vision and goals of the product. This high-level planning provides the context within which all subsequent planning activities occur. At the tactical level, stakeholders are involved in setting priorities and making trade-offs to maximize the product's value. At the operational level, their feedback helps refine the day-to-day activities and decisions of the development team.

To optimize stakeholder involvement, it is crucial to establish clear communication channels and feedback loops. Tools such as dashboards, progress reports, and regular updates can help keep stakeholders informed and engaged. Additionally, leveraging collaborative platforms and technologies can facilitate real-time communication and decision-making, especially in distributed or remote teams.

Challenges in stakeholder involvement often arise from misaligned expectations, conflicting priorities, and communication barriers. Addressing these challenges requires a proactive approach, including setting clear expectations from the outset, fostering a collaborative culture, and continuously seeking feedback to identify and resolve issues promptly.

Effective stakeholder involvement is not a one-time effort but an ongoing process that evolves with the product and the organization. By systematically integrating stakeholders into the Scrum process and the Agile Planning Onion, organizations can enhance the quality, relevance, and success of their products.

Product Increment Planning

Product increment planning is a critical aspect within the broader framework of Scrum and the Agile Planning Onion. It serves as a systematic approach to defining and organizing the deliverable components that will contribute to the overall product vision. This subchapter delves into the methodologies, principles, and practices that guide product increment planning, ensuring alignment with both strategic objectives and operational capabilities.

The primary objective of product increment planning is to delineate the scope and deliverables for each sprint, ensuring that each increment adds tangible value to the product. This process begins with the identification of key features and functionalities that are prioritized based on stakeholder input, market demands, and technical feasibility. The product backlog serves as the foundational repository from which these items are selected, with each item being meticulously refined and estimated to provide a clear understanding of the effort required.

Central to this planning process is the concept of the Definition of Done (DoD), a set of criteria that determines whether a product increment is complete and meets the required standards. The DoD is collaboratively established by the Scrum team and is consistently applied to ensure uniform quality across increments. This criterion encompasses aspects such as code quality, testing, documentation, and integration, thereby providing a comprehensive framework for assessing completeness.

The Scrum team, comprising the Product Owner, Scrum Master, and Development Team, plays a pivotal role in product increment planning. The Product Owner prioritizes the backlog items, ensuring alignment with the product vision and stakeholder expectations. The Development Team, on the other hand, provides insights into the technical aspects and feasibility of implementing the selected backlog items within the sprint. The Scrum Master facilitates this process, ensuring that the team adheres to Scrum principles and practices.

During the sprint planning meeting, the team collaboratively selects the backlog items that will constitute the product increment for the upcoming sprint. This selection is based on the team's capacity, historical velocity, and the prioritization set by the Product Owner. Each selected item is then broken down into smaller tasks, which are estimated and assigned to team members. This granular approach ensures that the team has a clear and actionable plan for the sprint, with each member understanding their responsibilities and deliverables.

A critical aspect of product increment planning is the continuous feedback loop, which is facilitated by regular reviews and retrospectives. The sprint review provides an opportunity for the team to demonstrate the completed increment to stakeholders, gathering feedback and validating that the increment meets the desired objectives. The sprint retrospective, on the other hand, allows the team to reflect on their processes and identify areas for improvement. This iterative feedback mechanism ensures that the planning process is adaptive and responsive to changing requirements and insights.

Product increment planning also involves risk management and contingency planning. The team identifies potential risks that could impact the delivery of the increment and devises strategies to mitigate these risks. This proactive approach ensures that the team is prepared to address challenges and can maintain the momentum of the sprint.

Incorporating tools and techniques such as user stories, acceptance criteria, and task boards enhances the efficiency and effectiveness of product increment planning. User stories provide a narrative description of the desired functionality from the perspective of the end user, while acceptance criteria define the conditions that must be met for the story to be considered complete. Task boards, whether physical or digital, offer a visual representation of the sprint progress, enabling the team to track tasks and identify bottlenecks.

In essence, product increment planning is a meticulous and collaborative process that integrates strategic vision with operational execution. By adhering to Scrum principles and leveraging iterative feedback, the team ensures that each increment delivers incremental value, contributing to the overarching goal of continuous product improvement.

Chapter 5: Release Planning

Defining Release Goals

Release goals are instrumental in guiding Scrum teams towards a coherent and strategic direction. These goals serve as a bridge between high-level vision and actionable tasks, ensuring that the team remains aligned with the overarching objectives of the project. Establishing clear and attainable release goals is essential for maintaining focus, motivation, and coherence throughout the development process.

Release goals encapsulate the primary outcomes that a team aims to achieve within a specified timeframe, typically spanning several sprints. These goals should be Specific, Measurable, Achievable, Relevant, and Time-bound (SMART), aligning closely with the product vision and stakeholder expectations. By doing so, they provide a clear and shared understanding of what constitutes success for the release.

The formulation of release goals begins with a thorough understanding of the product vision and roadmap. The product vision outlines the long-term aspirations and the value proposition of the product, while the roadmap provides a high-level timeline of how these aspirations will be realized. Release goals should be derived from these strategic documents, ensuring that each release incrementally contributes to the realization of the product vision.

Stakeholder involvement is crucial in defining release goals. Engaging with stakeholders—including customers, users, and business leaders—helps ensure that the goals are aligned with market needs and business priorities. This collaborative approach not only fosters buy-in and support but also enhances the relevance and impact of the release goals.

Once the context is established, the next step involves translating the high-level vision into more granular and actionable goals. This process often requires breaking down the product roadmap into smaller, manageable pieces that can be delivered incrementally. Each release goal should represent a valuable increment of the product, delivering tangible benefits to users and stakeholders.

The prioritization of release goals is another critical aspect. Given the dynamic nature of software development and market conditions, it's essential to prioritize goals based on their value, feasibility, and urgency. Techniques such as MoSCoW (Must have, Should have, Could have, Won't have) prioritization or Weighted Shortest Job First (WSJF) can be employed to systematically evaluate and rank the goals.

Transparency and communication are vital throughout the goal-setting process. The defined release goals should be communicated clearly and consistently to all team members and stakeholders. This ensures that everyone involved has a shared understanding of the objectives and can work collaboratively towards achieving them. Regular review and adjustment of the goals, based on feedback and changing circumstances, are also necessary to maintain their relevance and attainability.

Measurement and tracking of progress towards release goals are imperative. Key performance indicators (KPIs) and metrics should be established to monitor the team's progress and the impact of the release. This data-driven approach enables informed decision-making and facilitates continuous improvement.

In essence, defining release goals is a foundational activity that underpins the success of Scrum projects. It requires a strategic alignment with the product vision, active stakeholder engagement, meticulous planning, and effective communication. By setting clear, prioritized, and measurable goals, Scrum teams can navigate the complexities of software development with greater clarity and purpose.

Estimating and Forecasting

Estimating and forecasting are fundamental practices within the Scrum framework and the Agile Planning Onion model. These practices aim to provide a structured approach to predict the effort, time, and resources required to deliver a project successfully. Estimation and forecasting are iterative processes that incorporate empirical data, historical performance, and continuous feedback to enhance accuracy and reliability.

In Scrum, estimation is often conducted using techniques such as story points, T-shirt sizing, and planning poker. Story points are a unit of measure that expresses the overall effort required to implement a product backlog item. This method allows team members to compare the relative complexity and size of tasks rather than focusing on absolute time. T-shirt sizing categorizes tasks into predefined sizes (e.g., small, medium, large) to streamline the estimation process. Planning poker, a consensus-based technique, involves team members independently estimating the effort required for a task and then discussing their estimates to reach a collective agreement.

The Agile Planning Onion model delineates multiple layers of planning, from strategic vision to daily tasks. Estimation and forecasting play a crucial role at each layer. At the strategic level, long-term forecasts are developed based on high-level estimates and historical data. These forecasts guide decision-making regarding resource allocation, budgeting, and timeline projections. At the release planning level, estimates are refined to align with the scope and objectives of each release cycle. This involves breaking down high-level features into smaller, manageable tasks and estimating their effort.

Sprint planning, a core component of the Scrum framework, requires precise estimates to determine the workload for each iteration. Teams utilize velocity, a measure of the amount of work completed in previous sprints, to forecast the capacity for upcoming sprints. Velocity provides a data-driven basis for estimating how much work the team can realistically complete, thereby enhancing predictability and reducing the risk of overcommitment.

Forecasting in Scrum involves projecting future progress based on current trends and empirical data. Burn-down and burn-up charts are visual tools commonly used to track progress and forecast completion dates. A burn-down chart depicts the remaining work over time, providing a clear visual representation of the team's progress toward the sprint goal. A burn-up chart, on the other hand, displays both the total scope and completed work, offering insights into scope changes and overall progress.

Continuous improvement is a hallmark of Agile and Scrum methodologies. Retrospectives, conducted at the end of each sprint, offer an opportunity for teams to reflect on their estimation and forecasting practices. By analyzing discrepancies between estimated and actual effort, teams can identify patterns, refine their estimation techniques, and improve future forecasts.

The integration of estimation and forecasting within the Agile Planning Onion and Scrum framework ensures that planning is both adaptive and predictive. By leveraging empirical data, collaborative techniques, and iterative feedback, teams can enhance their ability to deliver value predictably and efficiently. Effective estimation and forecasting not only contribute to the successful execution of projects but also foster a culture of continuous learning and improvement within Agile teams.

Release Management

Release management in Scrum involves orchestrating the deployment of increments of potentially shippable products to end-users. This process is critical in ensuring that the delivered software meets the desired standards of quality, functionality, and timeliness. The Agile Planning Onion framework provides a structured approach to planning, which is essential for effective release management. Within this framework, release planning is situated between iteration planning and product planning, providing a bridge between short-term sprints and the long-term product vision.

The primary objective of release management is to align the development team's output with stakeholder expectations and market needs. This alignment is achieved through meticulous planning, continuous integration, and iterative feedback loops. A release plan typically encompasses multiple iterations, each delivering incremental value. The plan outlines the scope, schedule, and resources required to achieve the release goals, ensuring that all team members have a clear understanding of their roles and responsibilities.

Key activities in release management include defining the release scope, establishing timelines, and coordinating dependencies. The release scope is derived from the product backlog, which is prioritized based on business value, risk, and dependencies. The Product Owner plays a pivotal role in articulating the vision and ensuring that the release plan aligns with the strategic objectives. Timelines are established by estimating the effort required for each user story, taking into account the team's velocity and capacity. Dependencies, both internal and external, are identified and managed to minimize risks and ensure smooth progression towards the release.

Continuous integration and automated testing are fundamental practices in Scrum, enabling frequent validation of the product increment. These practices facilitate early detection and resolution of defects, thereby enhancing the overall quality of the release. Automated deployment pipelines further streamline the release process, reducing manual intervention and minimizing the likelihood of errors. The use of feature toggles allows for the release of new functionality in a controlled manner, providing flexibility to enable or disable features based on feedback and readiness.

Stakeholder communication is a critical component of release management. Regular updates and demonstrations ensure that stakeholders are informed about the progress and any potential changes to the release plan. This transparency fosters trust and collaboration, enabling quick decision-making and adjustments when necessary. The Sprint Review meetings serve as a platform for showcasing the completed work and gathering feedback, which is then incorporated into subsequent iterations.

Risk management is integral to release management in Scrum. Potential risks are identified, assessed, and mitigated through proactive planning and continuous monitoring. Contingency plans are developed to address high-impact risks, ensuring that the release can proceed smoothly even in the face of unforeseen challenges. Retrospectives provide an opportunity for the team to reflect on the release process, identify areas for improvement, and implement corrective actions in future releases.

The Agile Planning Onion emphasizes the iterative and incremental nature of planning, which is reflected in the release management process. By breaking down the release into manageable iterations, the team can deliver value early and often, while maintaining the flexibility to adapt to changing requirements. This approach not only enhances the predictability and reliability of releases but also ensures that the delivered product aligns closely with user needs and market demands.

Effective release management in Scrum requires a combination of strategic planning, technical excellence, and robust communication. By leveraging the principles of the Agile Planning Onion, teams can navigate the complexities of release management, delivering high-quality software that meets stakeholder expectations and drives business success.

Tracking and Reporting

Effective tracking and reporting mechanisms are indispensable components within the Scrum framework, ensuring both transparency and accountability. These mechanisms provide a structured approach for monitoring progress, identifying impediments, and facilitating informed decision-making. In the context of the Agile Planning Onion, tracking and reporting are vital at multiple layers, from strategic planning to daily task execution.

At the strategic layer, tracking progress towards long-term goals is crucial. This is typically achieved through key performance indicators (KPIs) and objectives and key results (OKRs). KPIs offer quantifiable measures that reflect the effectiveness of the Scrum team's efforts in achieving overarching business objectives. OKRs, on the other hand, provide a more dynamic framework by setting ambitious goals alongside measurable outcomes. Regular review cycles for these metrics ensure alignment with the broader organizational vision and allow for timely course corrections.

In the tactical layer, the focus shifts to tracking the progress of individual projects or releases. The Scrum framework utilizes several artifacts and ceremonies to facilitate this. The Product Backlog, maintained by the Product Owner, is a dynamic list of work items prioritized based on value and business impact. Regular backlog refinement sessions ensure that the backlog items are well-defined and ordered appropriately.

Sprint Planning meetings mark the commencement of each sprint, where the team commits to a set of Product Backlog items they aim to complete within the sprint duration. The Sprint Backlog is derived from this meeting, detailing the tasks necessary to deliver the selected work items. This artifact serves as a real-time tracker of the team's progress throughout the sprint.

Daily Scrums, or stand-up meetings, are pivotal for day-to-day tracking. These brief sessions provide a platform for team members to share updates on their progress, highlight any impediments, and synchronize their efforts. The transparency fostered by Daily Scrums ensures that potential issues are identified early, allowing for prompt intervention.

Burndown charts are another crucial tool for tracking sprint progress. These visual representations plot the remaining work against the sprint timeline, offering a clear indication of whether the team is on track to meet their commitments. By analyzing the trend lines in a burndown chart, teams can identify patterns and take corrective actions as needed.

At the operational layer, detailed task tracking is essential. Tools such as Kanban boards or task boards provide a visual representation of the workflow, categorizing tasks into columns such as "To Do," "In Progress," and "Done." This visual management technique enhances the team's ability to monitor task status, identify bottlenecks, and optimize workflow efficiency.

Reporting mechanisms complement the tracking processes by providing structured insights into the team's performance. Sprint Reviews and Retrospectives are integral ceremonies in this regard. During the Sprint Review, the team demonstrates the completed work to stakeholders, gathering feedback and validating the increment against the sprint goals. This ceremony not only serves as a progress report but also fosters stakeholder engagement and alignment.

The Sprint Retrospective, conducted at the end of each sprint, focuses on introspection and continuous improvement. The team reflects on their processes, identifying strengths, areas for improvement, and actionable steps to enhance their performance in subsequent sprints. This iterative approach to process refinement is a cornerstone of the Agile philosophy.

In summary, effective tracking and reporting within the Scrum framework are multi-faceted, encompassing strategic, tactical, and operational layers. Through a combination of artifacts, ceremonies, and visual management tools, Scrum teams can maintain transparency, ensure accountability, and drive continuous improvement. These mechanisms not only facilitate the successful delivery of work items but also align the team's efforts with the broader organizational goals, thereby maximizing value delivery.

Chapter 6: Iteration Planning

Sprint Planning Basics

Sprint Planning is a foundational activity in Scrum, serving as the initial step of the Sprint cycle. It is designed to establish a clear and achievable set of objectives for the upcoming Sprint, ensuring that the development team, Product Owner, and stakeholders are aligned on priorities and deliverables. This process is integral to the Agile Planning Onion, which conceptualizes planning as a layered, iterative activity.

The Sprint Planning meeting is typically time-boxed to a maximum of eight hours for a one-month Sprint, with shorter Sprints proportionally reducing the duration. The meeting involves the entire Scrum Team: the Product Owner, the Scrum Master, and the Development Team. Each role has distinct responsibilities that contribute to the effectiveness of the planning session.

The Product Owner plays a crucial role by presenting the prioritized Product Backlog items to the team. These items should be well-defined, refined, and estimated to facilitate informed decision-making. The Product Owner must ensure that the backlog items are aligned with the product vision and stakeholder expectations. During Sprint Planning, the Product Owner answers questions and provides clarifications to help the Development Team understand the scope and context of the work.

The Development Team is responsible for assessing the feasibility of the proposed backlog items. They engage in detailed discussions to break down the items into smaller, actionable tasks. This process often involves collaborative estimation techniques, such as Planning Poker, to gauge the effort required for each task. The team commits to a set of Product Backlog items they believe can be completed within the Sprint, considering their capacity and any known impediments.

The Scrum Master facilitates the Sprint Planning meeting, ensuring that the discussion remains focused and productive. The Scrum Master also helps the team adhere to Scrum principles and practices, addressing any impediments that may arise during the planning process. By fostering a collaborative environment, the Scrum Master enables the team to reach a consensus on the Sprint Goal and the Sprint Backlog.

The outcome of Sprint Planning is twofold: the Sprint Goal and the Sprint Backlog. The Sprint Goal is a concise statement that defines the overarching objective of the Sprint. It provides a clear direction and purpose, guiding the team's efforts and decision-making throughout the Sprint. The Sprint Backlog, on the other hand, is a detailed list of tasks and deliverables that the team commits to completing. It includes the selected Product Backlog items and the tasks necessary to achieve the Sprint Goal.

Effective Sprint Planning requires thorough preparation and active participation from all team members. The Product Owner should ensure that the Product Backlog is well-groomed and prioritized. The Development Team should come prepared with an understanding of their capacity and any potential impediments. The Scrum Master should facilitate the meeting with a focus on maintaining adherence to Scrum principles and fostering collaboration.

Sprint Planning is not a static process; it evolves with the team's maturity and experience. Continuous improvement practices, such as retrospectives, help the team refine their planning techniques and enhance their effectiveness. By iteratively refining their approach, the team can achieve greater alignment, predictability, and value delivery.

In summary, Sprint Planning is a critical activity that sets the stage for a successful Sprint. It aligns the team on priorities, establishes a clear direction, and ensures a shared understanding of the work ahead. Through collaborative effort and adherence to Scrum principles, the team can effectively plan and execute their Sprints, delivering valuable increments of the product.

Task Breakdown and Assignment

In the context of Scrum and the Agile Planning Onion, breaking down tasks and assigning them effectively are pivotal components for achieving project success. This subchapter elucidates the methodologies and principles underlying task decomposition and the subsequent assignment of responsibilities within an Agile team.

The initial phase involves decomposing user stories or high-level requirements into smaller, manageable tasks. This granular breakdown is essential for creating a clear and actionable roadmap. The process typically begins during the sprint planning meeting, where the Product Owner and Scrum Team collaboratively refine the backlog. Each user story is dissected into constituent tasks that can be completed within a single sprint. This decomposition not only enhances understanding but also facilitates accurate estimation and prioritization.

The MoSCoW method (Must have, Should have, Could have, and Won't have) often guides prioritization. Tasks categorized as 'Must have' are indispensable for the sprint goal, whereas 'Should have' and 'Could have' tasks are secondary in importance. This prioritization ensures that critical functionalities are addressed first, aligning with the Agile principle of delivering maximum value early and often.

Estimations are conducted using techniques such as Planning Poker or the Fibonacci sequence. These collaborative estimation methods leverage collective team insights to gauge the effort required for each task. The consensus-driven approach mitigates biases and results in more accurate estimations. The use of story points rather than hours is advocated to focus on relative effort and complexity, fostering a more adaptive and flexible planning process.

Once tasks are clearly defined and estimated, the next step is their assignment. Effective task assignment hinges on understanding team members' skills, expertise, and current workload. The Scrum Master plays a crucial role in facilitating this process, ensuring that tasks are distributed equitably and align with individual strengths. Cross-functional teams are a hallmark of Agile, promoting a culture where team members are encouraged to acquire new skills and collaborate beyond their primary expertise areas.

The concept of self-organization is fundamental in Agile task assignment. Team members often volunteer for tasks based on their interests and capabilities, fostering a sense of ownership and accountability. This approach contrasts with traditional top-down assignment methods, empowering teams to take proactive roles in their work. The Scrum Master supports this self-organizing behavior by removing impediments and ensuring that the team remains focused on the sprint goals.

Daily stand-up meetings, or daily Scrums, provide a platform for continuous task reassignment as needed. These brief, time-boxed meetings allow team members to share progress, identify obstacles, and adjust assignments dynamically. This iterative reassignment ensures that the team remains agile and responsive to changing circumstances, maintaining momentum towards sprint completion.

Moreover, visual management tools like Kanban boards or digital equivalents (e.g., JIRA, Trello) are instrumental in tracking task progress. These tools provide real-time visibility into task status, fostering transparency and collaboration. Each task moves through stages such as 'To Do,' 'In Progress,' and 'Done,' enabling the team to monitor workflow and identify bottlenecks promptly.

Incorporating feedback loops through regular retrospectives ensures continuous improvement in task breakdown and assignment processes. These sessions allow the team to reflect on what worked well and identify areas for enhancement. By iterating on their approach, the team can refine their strategies for task decomposition and assignment, driving towards higher efficiency and effectiveness in future sprints.

This systematic approach to task breakdown and assignment, grounded in Agile principles, ensures that teams remain focused, adaptable, and aligned with project goals, ultimately contributing to the successful delivery of high-quality products.

Daily Stand-ups

Daily stand-ups are an integral component of the Scrum framework, serving as a cornerstone for facilitating communication, fostering collaboration, and ensuring alignment among team members. These brief, time-boxed meetings are designed to enhance transparency and adaptability, which are pivotal principles of Agile methodologies. The stand-up is typically conducted daily and should not exceed 15 minutes, ensuring it remains efficient and focused.

The primary objective of the daily stand-up is to synchronize the team's activities and identify any impediments that might hinder progress. Each team member is expected to address three key questions: (1) What did I accomplish yesterday? (2) What will I work on today? (3) Are there any obstacles impeding my progress? By systematically addressing these queries, the team can maintain a clear understanding of the current state of the project and make informed decisions to navigate challenges.

The facilitation of daily stand-ups requires adherence to specific guidelines to maximize their effectiveness. Firstly, it is crucial that the stand-up occurs at the same time and place each day, establishing a routine that reinforces discipline and punctuality. This consistency aids in embedding the practice into the team's daily workflow, minimizing disruptions and ensuring that all members can participate without scheduling conflicts.

Furthermore, the stand-up should be conducted in a standing position, as the physical posture encourages brevity and focus. This practice helps to prevent the meeting from devolving into a prolonged discussion, maintaining the intended time-boxed nature.

The role of the Scrum Master in daily stand-ups is to act as a facilitator rather than a directive authority. The Scrum Master ensures that the meeting adheres to its time constraints and agenda while fostering an environment where team members feel comfortable sharing their updates and concerns. If any significant issues or detailed discussions arise, the Scrum Master should take note and address them in a separate, more appropriate forum to prevent the stand-up from being derailed.

Effective daily stand-ups also hinge on the active participation and engagement of all team members. Each participant should come prepared to provide concise and relevant updates. This preparation not only respects the time of others but also enhances the quality of the information shared, leading to more accurate and timely decision-making.

Technological tools can augment the efficacy of daily stand-ups, particularly in distributed teams. Video conferencing platforms, shared digital boards, and collaboration software can bridge geographical gaps and ensure that all team members, regardless of location, can contribute effectively. Such tools can also facilitate the documentation and tracking of action items and impediments, providing a transparent and accessible record for the team.

The impact of daily stand-ups on team dynamics and project outcomes is significant. By fostering regular communication, these meetings help to build a culture of accountability and collective ownership. They enable the early detection of issues, allowing for prompt intervention and course correction. Moreover, the iterative nature of the stand-up aligns with the Agile philosophy of continuous improvement, as it provides a daily opportunity for reflection and adjustment.

In essence, daily stand-ups are a vital practice within the Scrum framework, embodying the principles of transparency, inspection, and adaptation. Their structured yet flexible nature supports the dynamic needs of Agile teams, contributing to the overall success and resilience of the project.

Review and Retrospective

The efficacy of Scrum as an Agile framework hinges on its iterative nature, which is fundamentally underscored by the Review and Retrospective ceremonies. These ceremonies serve as pivotal moments for teams to assess their progress, adapt processes, and foster continuous improvement.

The Review, typically conducted at the end of each sprint, provides an opportunity for the Scrum team to showcase the increment of work completed. This demonstration is not merely a presentation but an interactive session where feedback from stakeholders is solicited. The primary objective is to inspect the product increment and adapt the product backlog based on this feedback. The Review ensures alignment with stakeholder expectations and promotes transparency. It is essential that the Review is facilitated in a manner that encourages open dialogue, enabling stakeholders to provide constructive feedback that can be used to refine future work.

Quantitative metrics such as velocity, burndown charts, and work item flow are often discussed during the Review to provide a data-driven assessment of the team's performance. However, it is equally important to consider qualitative feedback, which can offer insights into customer satisfaction and usability that metrics alone cannot capture. The Review should culminate in actionable items that inform the next sprint planning session, ensuring that the team remains responsive to changing requirements and stakeholder needs.

Following the Review, the Retrospective is conducted as the final event of the sprint. The Retrospective is an introspective session focused on the team's processes rather than the product. Its purpose is to identify what went well, what could be improved, and how the team can enhance its workflow. The Retrospective is a critical component of the Scrum framework as it fosters a culture of continuous improvement and collective ownership of the process.

During the Retrospective, the team employs various techniques and tools to facilitate discussion, such as the Start-Stop-Continue method, fishbone diagrams, or the Five Whys technique. These methods help the team to systematically identify root causes of any issues encountered during the sprint and propose solutions. It is crucial that the Retrospective environment is safe and non-judgmental, encouraging team members to speak freely about their experiences. Psychological safety is paramount; without it, the effectiveness of the Retrospective can be severely compromised.

The outcomes of the Retrospective should include specific, measurable actions that the team commits to implementing in the upcoming sprints. These actions are often captured in an improvement backlog, which is reviewed and updated regularly. The commitment to these actions demonstrates the team's dedication to refining their processes and enhancing their performance continuously.

The cyclical nature of the Review and Retrospective ensures that Scrum teams remain adaptive and resilient. By regularly reflecting on their work and processes, teams can identify and address issues before they escalate, thereby maintaining a steady trajectory of improvement. The Review and Retrospective are not isolated events but integral components of a feedback loop that drives the Scrum framework.

In essence, the Review and Retrospective ceremonies are vital mechanisms for ensuring that both the product and the process are continuously evolving to meet the needs of stakeholders and the team. They embody the principles of transparency, inspection, and adaptation, which are the cornerstones of Scrum and the broader Agile methodology. By diligently conducting these ceremonies, Scrum teams can achieve a dynamic equilibrium between delivering value and optimizing their workflows.

Chapter 7: The Role of Scrum Master

Facilitating Scrum Events

Scrum events are integral to the effective implementation of the Scrum framework, serving as structured opportunities for teams to inspect and adapt their processes. The facilitation of these events is a critical function that ensures their success and maximizes their value. Effective facilitation requires a nuanced understanding of both the theoretical underpinnings of Scrum and the practical challenges teams face.

Scrum events, including the Sprint Planning, Daily Scrum, Sprint Review, and Sprint Retrospective, each have distinct purposes and structures. The facilitator's role in these events is to guide the team in adhering to Scrum principles while fostering an environment conducive to collaboration and continuous improvement.

The Sprint Planning meeting sets the stage for the entire sprint. During this event, the Scrum Team collaborates to define the sprint goal and select items from the product backlog to work on. The facilitator's responsibility is to ensure that the team remains focused on the objective, encourages participation from all members, and helps the team to estimate and select the appropriate amount of work. Effective facilitation here involves balancing the input from the Product Owner, who provides the vision and priorities, with the Development Team's capacity and technical considerations.

The Daily Scrum, often referred to as the stand-up meeting, is a short, time-boxed event where team members synchronize their activities and create a plan for the next 24 hours. The facilitator helps maintain the meeting's brevity and focus, ensuring it does not devolve into a problem-solving session. By encouraging concise updates and fostering a sense of accountability, the facilitator helps the team to stay aligned and identify impediments early.

The Sprint Review is an opportunity for the Scrum Team to present the increment of work completed during the sprint to stakeholders. The facilitator's role is to create a collaborative atmosphere where feedback is openly shared and the increment is inspected. This requires skill in managing diverse stakeholder perspectives and ensuring that the discussion remains constructive and aligned with the sprint goals.

The Sprint Retrospective, the final event of the sprint, focuses on the team's process and practice improvements. The facilitator encourages honest reflection and the identification of actionable improvements. This involves creating a safe space for team members to express their thoughts and fostering a culture of continuous improvement. Techniques such as root cause analysis and brainstorming can be employed to uncover underlying issues and generate solutions.

A proficient facilitator adapts their approach based on the team's maturity and the specific context of each event. They utilize various techniques and tools to enhance engagement and ensure that each event fulfills its intended purpose. For example, the use of visual aids, time-boxing strategies, and collaborative tools can help maintain focus and drive effective outcomes.

Facilitating Scrum events is not merely about following a prescribed set of steps but about understanding the dynamics of the team and guiding them towards self-organization and improvement. It demands a blend of skills, including active listening, conflict resolution, and the ability to ask powerful questions. By mastering these skills, a facilitator can significantly enhance the effectiveness of Scrum events, contributing to the overall success of the Agile planning process.

Coaching the Team

Effective team coaching within the framework of Scrum and the Agile Planning Onion is a multifaceted endeavor that necessitates a deep understanding of both individual and collective team dynamics. The primary objective of coaching in this context is to facilitate the team's ability to self-organize, enhance performance, and achieve continuous improvement. This involves not only imparting technical skills and knowledge but also fostering an environment where team members can collaborate effectively and adapt to changing circumstances.

A key aspect of coaching a Scrum team is the establishment and maintenance of psychological safety. Psychological safety, a term coined by Amy Edmondson, refers to a shared belief that the team is safe for interpersonal risk-taking. In a psychologically safe environment, team members feel comfortable expressing their thoughts, asking questions, and admitting mistakes without fear of ridicule or retribution. This is crucial for fostering open communication and collaboration, which are foundational to the success of Scrum and Agile methodologies.

The role of the coach also involves guiding the team through the various stages of group development, as described by Tuckman's model: forming, storming, norming, and performing. During the forming stage, the coach helps the team members get to know each other and understand their roles within the team. In the storming stage, the coach assists in resolving conflicts and aligning the team towards common goals. As the team progresses to the norming stage, the coach supports the establishment of effective work processes and collaboration norms. Finally, in the performing stage, the coach encourages the team to strive for high performance and continuous improvement.

Another critical component of coaching is facilitating effective Scrum ceremonies, including sprint planning, daily stand-ups, sprint reviews, and retrospectives. During sprint planning, the coach ensures that the team clearly understands the sprint goals and the work required to achieve them. In daily stand-ups, the coach helps the team maintain focus and address any impediments swiftly. During sprint reviews, the coach encourages constructive feedback and collaboration with stakeholders. In retrospectives, the coach guides the team in reflecting on their performance and identifying actionable improvements.

Coaching also involves promoting the principles of the Agile Planning Onion, which emphasizes the iterative and incremental approach to planning. The layers of the Agile Planning Onion—vision, roadmap, release, iteration, and daily planning—serve as a guide for the team to align their efforts with the broader organizational goals while remaining flexible to adapt to changes. The coach plays a pivotal role in helping the team navigate these layers, ensuring that planning is conducted at the appropriate level of detail and that the team remains focused on delivering value.

In addition to these responsibilities, the coach must also provide individual support to team members. This may involve offering mentorship, facilitating skill development, and providing constructive feedback. By addressing individual needs, the coach helps to build a stronger, more cohesive team.

Ultimately, the effectiveness of team coaching in Scrum and Agile methodologies hinges on the coach's ability to balance guidance with empowerment. The coach must provide the necessary support and direction while encouraging the team to take ownership of their processes and outcomes. By fostering a culture of collaboration, continuous learning, and adaptability, the coach helps the team to not only meet but exceed their goals.

Removing Impediments

Impediments in the Scrum framework are obstacles that hinder a team's progress towards achieving their sprint goals. Identifying and removing these impediments is crucial to maintaining the efficiency and effectiveness of the Scrum process. The responsibility of removing impediments primarily falls on the Scrum Master, whose role includes facilitating the team's ability to reach optimal productivity.

Impediments can arise from various sources, including organizational, technical, and team-related issues. Organizational impediments might include hierarchical structures or bureaucratic processes that slow down decision-making. Technical impediments could involve outdated tools, insufficient infrastructure, or a lack of access to necessary resources. Team-related impediments might include interpersonal conflicts, skill gaps, or unclear roles and responsibilities.

The first step in addressing impediments is identification. This is achieved through continuous monitoring and communication within the team. Daily stand-ups, sprint reviews, and retrospectives provide structured opportunities for team members to voice concerns and highlight obstacles. The Scrum Master must cultivate an environment where team members feel comfortable raising issues without fear of blame or retribution.

Once identified, the next step is to analyze the impediments to understand their root causes. This often involves collaboration with stakeholders outside the Scrum team, such as managers, other teams, or external partners. Techniques such as the "Five Whys" or root cause analysis can be employed to dig deeper into the underlying issues. Understanding the root cause is essential for developing effective solutions rather than merely addressing symptoms.

The Scrum Master then takes proactive measures to remove the impediments. Organizational impediments may require negotiating with higher management to streamline processes or secure additional resources. For technical impediments, the Scrum Master might coordinate with IT departments or external vendors to upgrade tools or infrastructure. Addressing team-related impediments could involve facilitating conflict resolution sessions, arranging training programs, or clarifying roles and responsibilities.

It is important to note that some impediments may not be immediately solvable. In such cases, the Scrum Master must prioritize these impediments based on their impact on the team's productivity and work on mitigating their effects until a permanent solution is found. This could involve implementing temporary workarounds or reallocating team resources to minimize disruption.

Continuous improvement is a fundamental principle of Scrum, and removing impediments is an ongoing process. The Scrum Master should regularly review the effectiveness of the actions taken to address impediments and adjust strategies as necessary.

This iterative approach ensures that the team remains agile and can adapt to new challenges as they arise.

Metrics and feedback loops are essential tools in this process. By measuring the impact of impediments and the effectiveness of solutions, the Scrum Master can make data-driven decisions. Key performance indicators (KPIs) such as velocity, cycle time, and team satisfaction can provide valuable insights into the team's health and highlight areas that require attention.

Effective communication and collaboration are critical throughout this process. The Scrum Master must maintain open lines of communication with all stakeholders and ensure that everyone is aligned on the priorities and progress of impediment removal efforts. This transparency fosters a culture of trust and collective responsibility, which is essential for the successful implementation of Scrum.

In essence, the removal of impediments is a dynamic and collaborative effort that requires vigilance, analysis, and proactive intervention. By systematically identifying, addressing, and reviewing impediments, the Scrum Master plays a pivotal role in enabling the Scrum team to achieve their goals and deliver value continuously.

Continuous Improvement

Continuous improvement is a fundamental principle in Agile methodologies, particularly within the Scrum framework. This iterative approach to enhancing processes, products, and services ensures that teams do not become complacent but rather remain adaptive and responsive to changing conditions. The Agile Planning Onion, a conceptual model that layers planning activities from strategic to tactical levels, incorporates continuous improvement at every layer, ensuring that feedback loops are integral to the process.

In Scrum, the Sprint Review and Sprint Retrospective are critical ceremonies that facilitate continuous improvement. The Sprint Review focuses on the product, involving stakeholders in the evaluation of the increment to gather feedback and adapt the product backlog accordingly. This iterative inspection allows the team to align the product increment with the stakeholders' expectations and market demands, ensuring that the product evolves in the right direction.

The Sprint Retrospective, on the other hand, targets the process and team dynamics. During this ceremony, the Scrum Team reflects on the past sprint to identify successes, areas for improvement, and actionable steps to enhance their way of working. This introspection is crucial for fostering a culture of openness and accountability, where team members feel empowered to voice their opinions and suggest improvements.

By regularly examining their processes and interactions, teams can optimize their workflows, reduce waste, and enhance overall efficiency.

Metrics and data play a pivotal role in driving continuous improvement within Scrum. Key performance indicators (KPIs) such as velocity, cycle time, and defect rates provide quantitative insights into the team's performance. These metrics, when analyzed over multiple sprints, reveal trends and patterns that can inform decisions on process adjustments. For example, a consistent decline in velocity may indicate underlying issues such as technical debt or impediments, prompting the team to investigate and address the root causes.

Furthermore, continuous improvement extends beyond the confines of individual Scrum teams. Organizations adopting Scrum at scale, such as through the Scaled Agile Framework (SAFe) or Large-Scale Scrum (LeSS), must ensure that feedback loops exist at all levels. Program Increment (PI) Planning and Inspect & Adapt (I&A) workshops in SAFe, for instance, provide opportunities for cross-team and cross-functional alignment, enabling the organization to identify systemic issues and implement improvements that benefit the entire portfolio.

The role of leadership in fostering a culture of continuous improvement cannot be understated. Leaders must create an environment that encourages experimentation, learning from failures, and celebrating successes. Psychological safety is paramount, as it allows team members to take risks and innovate without fear of retribution. Leaders should also invest in ongoing training and development, ensuring that teams are equipped with the latest tools, techniques, and knowledge to enhance their performance.

Moreover, the integration of modern tools and technologies can significantly augment continuous improvement efforts. Agile project management tools such as Jira, Trello, and Azure DevOps offer features that facilitate tracking, reporting, and analyzing metrics. These tools also support automation of repetitive tasks, freeing up team members to focus on value-added activities. Continuous Integration/Continuous Deployment (CI/CD) pipelines, for example, enable faster feedback loops by automating testing and deployment processes, thus accelerating the delivery of high-quality increments.

In summary, continuous improvement within the Scrum framework and the Agile Planning Onion is a multifaceted endeavor that encompasses iterative feedback, data-driven decision-making, leadership support, and technological advancements. By embedding continuous improvement into the fabric of the organization, teams can achieve sustained agility, innovation, and competitive advantage.

Chapter 8: The Role of Product Owner

Defining the Product Vision

The foundation of any successful product development initiative lies in the establishment of a clear and compelling product vision. This vision serves as the guiding star for all subsequent planning and execution activities within the Agile framework. It is imperative that the product vision is both aspirational and achievable, providing a coherent direction for the team and aligning stakeholders with the overarching goals.

A product vision encapsulates the essence of what the product aims to achieve and the value it intends to deliver to its users. It is not merely a statement of intent but a strategic tool that influences decision-making at every level of the project. The vision must articulate the problem the product seeks to solve, the target audience, and the unique value proposition that differentiates it from competitors.

To craft an effective product vision, it is crucial to engage in a thorough understanding of the market landscape and user needs. This involves conducting comprehensive market research, user interviews, and competitive analysis. The insights gleaned from these activities inform the vision, ensuring it is grounded in real-world requirements and opportunities.

The product vision should be succinct yet comprehensive. A well-formulated vision statement typically comprises the following elements:

1. **Purpose**: Clearly stating the fundamental reason for the product's existence. This is often articulated in terms of the problem it solves or the opportunity it capitalizes on.

2. **Scope**: Defining the boundaries within which the product will operate. This includes specifying the primary features and functionalities that the product will offer.

3. **Target Audience**: Identifying the primary users or customers of the product. Understanding the demographics, behaviors, and needs of this group is critical for tailoring the product to their requirements.

4. **Differentiation**: Highlighting what sets the product apart from existing alternatives. This could be in terms of unique features, superior user experience, cost advantages, or other distinguishing factors.

5. **Long-term Goals**: Outlining the strategic objectives the product aims to achieve over its lifecycle. These goals should be

ambitious yet realistic, providing a roadmap for future development and growth.

Effective communication of the product vision is as important as its formulation. It should be shared and endorsed by all stakeholders, including team members, management, and external partners. This ensures a unified understanding and collective commitment to the vision. Visual aids, such as vision boards or infographics, can be employed to make the vision more tangible and engaging.

The product vision also plays a pivotal role in the Agile planning process. It acts as the topmost layer of the Agile Planning Onion, influencing all subsequent layers, including the roadmap, release planning, iteration planning, and daily planning. By maintaining a clear focus on the vision, teams can ensure that their efforts are consistently aligned with the strategic objectives.

It is important to recognize that the product vision is not static. As market conditions evolve and new insights emerge, the vision may need to be revisited and refined. This iterative approach ensures that the product remains relevant and continues to meet the needs of its users.

Establishing a robust product vision is a critical step in the Agile planning process. It provides a strategic framework that guides decision-making, fosters alignment among stakeholders, and ensures that the product delivers meaningful value. As such, it is a fundamental component of successful product development within the Scrum methodology.

Managing the Product Backlog

The Product Backlog is a dynamic and evolving artifact within the Scrum framework, serving as the single source of truth for all the work that needs to be done. Effective management of the Product Backlog is crucial to ensure that the Scrum team can deliver maximum value to stakeholders. This involves several key practices and principles that align with the iterative and incremental nature of Scrum.

At the core of managing the Product Backlog is the role of the Product Owner. The Product Owner is responsible for maintaining the backlog, ensuring it is visible, transparent, and understood by all members of the team. The backlog must be ordered, with the most valuable and highest-priority items at the top, allowing the team to focus on delivering the most important features first. This ordering is not static; it requires continuous refinement and adjustment based on feedback and changing circumstances.

Refinement of the Product Backlog is an ongoing activity where the Product Owner collaborates with the Development Team to break down larger items into smaller, more manageable ones. This process, often referred to as backlog grooming, helps in clarifying requirements, estimating effort, and ensuring that the backlog items are ready for future sprints. Effective refinement sessions involve detailed discussions, acceptance criteria definition, and sometimes revisiting the user stories to align them with the current understanding of the product's needs.

One of the fundamental principles in managing the Product Backlog is transparency. All stakeholders should have a clear understanding of the backlog's content and priorities. This transparency is achieved through regular communication and documentation, ensuring that the backlog reflects the latest information and decisions. Tools such as digital backlog management systems can aid in maintaining this transparency, providing a centralized platform where all changes and updates are tracked and visible.

Prioritization is another critical aspect of backlog management. The Product Owner must weigh various factors such as business value, risk, dependencies, and technical feasibility when ordering the backlog. Techniques like the MoSCoW method (Must have, Should have, Could have, and Won't have), Kano model, and cost of delay can be employed to prioritize backlog items effectively. It is essential to engage stakeholders during this process to ensure that their needs and expectations are adequately represented and addressed.

Quality is a non-negotiable aspect of the Product Backlog. Each item should be well-defined, with clear acceptance criteria that outline the conditions under which the work will be considered complete. This clarity helps the Development Team in understanding what is expected and reduces ambiguity, leading to higher quality deliverables. Regularly reviewing and updating acceptance criteria based on feedback and changes in requirements ensures that the backlog remains relevant and actionable.

The Product Backlog is not just a list of tasks but a strategic tool that guides the development of the product. It should align with the product vision and roadmap, ensuring that every item contributes to the overall goals and objectives. Regularly revisiting the product vision and roadmap during backlog refinement sessions helps in maintaining this alignment and making informed decisions about backlog priorities.

Effective management of the Product Backlog requires a balance of strategic planning and tactical execution. It is an ongoing process that demands active involvement from the Product Owner, collaboration with the Development Team, and constant communication with stakeholders. By adhering to these principles and practices, the Scrum team can ensure that the backlog remains a valuable and actionable artifact, driving the successful delivery of the product.

Prioritizing Work

Effective prioritization within the Scrum framework is pivotal for optimizing productivity and achieving project goals. Prioritization ensures that the most valuable and critical tasks are addressed first, aligning with both stakeholder expectations and team capacity.

The Product Backlog serves as the primary source of work items, containing features, enhancements, and bug fixes. Prioritization within the Product Backlog is conducted by the Product Owner, who is responsible for maximizing the value of the product. This involves evaluating each backlog item based on several criteria, including business value, risk, dependencies, and the effort required for implementation. The MoSCoW method (Must have, Should have, Could have, and Won't have) is often employed to categorize tasks effectively. Must-have items are essential for the product's viability, while should-have and could-have items enhance the product but are not critical. Won't-have items are deprioritized for the current planning cycle.

Value-driven prioritization is a core principle. The Product Owner must constantly communicate with stakeholders to understand their needs and expectations. This dialogue ensures that the backlog reflects the most current and relevant requirements. Value is typically assessed in terms of return on investment (ROI), customer satisfaction, and strategic alignment with business objectives. Higher-value items are prioritized to deliver maximum impact early in the development cycle.

Risk management also plays a crucial role in prioritization. High-risk items, which may have significant unknowns or technical challenges, are often tackled early to mitigate potential issues. This proactive approach allows the team to address uncertainties and adapt the plan based on findings, reducing the likelihood of major disruptions later in the project.

Dependencies between tasks must be carefully managed. Items that unlock other backlog entries or are prerequisites for subsequent work are given higher priority. This sequencing ensures a smooth workflow and avoids bottlenecks. The Product Owner, in collaboration with the development team, identifies and plans for these dependencies during backlog refinement sessions.

Effort estimation is another critical factor. The Scrum team uses techniques such as Planning Poker or T-shirt sizing to estimate the effort required for each backlog item. These estimates help the Product Owner prioritize tasks that can be realistically completed within the upcoming Sprint. Smaller, well-defined tasks are often prioritized over larger, ambiguous ones to maintain momentum and ensure continuous delivery of value.

The concept of technical debt must also be considered. Technical debt refers to the implied cost of additional rework caused by choosing an easy solution now instead of a better approach that would take longer. Balancing the need for new features with the necessity of addressing technical debt is essential. Neglecting technical debt can lead to a decrease in productivity and an increase in maintenance costs over time.

Prioritization is not a one-time activity but an ongoing process. The Product Owner revisits and reorders the backlog regularly, incorporating feedback from stakeholders, market changes, and the development team's insights. This dynamic approach ensures that the team remains focused on the most valuable and relevant work items.

Scrum's iterative nature inherently supports continuous prioritization. At the end of each Sprint, during the Sprint Review, the team and stakeholders assess the progress and make necessary adjustments to the backlog. This feedback loop allows for real-time prioritization adjustments based on the latest information and project status.

Effective prioritization within the Scrum framework is essential for delivering high-value products efficiently. By systematically evaluating business value, managing risks, addressing dependencies, and considering effort estimates and technical debt, the Product Owner ensures that the Scrum team focuses on the most impactful work. This disciplined approach to prioritization enables the team to deliver incremental value consistently, meeting stakeholder expectations and achieving project objectives.

Collaborating with Stakeholders

Effective collaboration with stakeholders is a critical component in the successful implementation of Scrum and Agile methodologies. Stakeholders encompass a broad range of individuals and groups, including customers, end-users, management, and team members who have a vested interest in the outcome of the project. The Agile Planning Onion, an iterative model that outlines the levels of planning involved in Agile projects, places significant emphasis on stakeholder engagement at all stages.

The initial layer of the Agile Planning Onion, the product vision, necessitates direct input from stakeholders. This vision represents the overarching goal and purpose of the project, setting the stage for all subsequent planning activities. Stakeholders provide insights into market needs, business objectives, and customer expectations, which are crucial for formulating a coherent and compelling product vision. Regular consultations with stakeholders ensure that the vision remains aligned with their evolving needs and market dynamics.

Progressing to the next layer, the product roadmap, stakeholder collaboration becomes even more pronounced. The roadmap outlines the high-level plan for achieving the product vision, including major milestones and deliverables. Stakeholders contribute to defining these milestones, prioritizing features, and identifying potential risks. Their input helps in balancing technical feasibility with business value, thereby ensuring that the roadmap is both realistic and strategically sound.

At the release planning level, stakeholders play a pivotal role in determining the scope and timing of product releases. This involves negotiating trade-offs between scope, time, and resources to achieve the desired balance of value delivery and risk management. Stakeholders' feedback on release plans helps in adjusting priorities and refining deliverables to better meet market demands and organizational goals. Their involvement ensures that releases are not only technically viable but also commercially advantageous.

Sprint planning, a more granular level within the Agile Planning Onion, also benefits from stakeholder collaboration. During sprint planning meetings, the development team works closely with the product owner and other stakeholders to select and prioritize user stories or tasks for the upcoming sprint. Stakeholders provide context and clarification on requirements, helping the team to understand the underlying business value and user needs. This collaboration enhances the team's ability to deliver increments of value that are closely aligned with stakeholder expectations.

The daily scrum meetings, while primarily focused on team coordination, also provide an opportunity for stakeholders to stay informed about progress and impediments. Although stakeholders do not typically participate in these meetings, their presence can be beneficial during critical junctures or when significant issues arise. Regular updates to stakeholders help in maintaining transparency and fostering trust, which are essential for effective collaboration.

Review and retrospective meetings offer formal avenues for stakeholders to provide feedback on completed work and the overall process. In review meetings, stakeholders assess the delivered increments and provide constructive feedback, which is crucial for continuous improvement. Retrospective meetings, on the other hand, focus on identifying process improvements and addressing any collaboration challenges. Stakeholder participation in these meetings ensures that their perspectives are considered in both product and process enhancements.

The iterative nature of Agile and Scrum necessitates ongoing engagement with stakeholders at all levels of the planning onion. Effective collaboration leads to better alignment between the development team and stakeholders, resulting in products that not only meet technical specifications but also deliver significant business value. By fostering a culture of open communication and mutual respect, teams can leverage stakeholder expertise and insights to navigate the complexities of Agile projects successfully.

Chapter 9: The Development Team

Self-Organizing Teams

The concept of self-organizing teams is a cornerstone of Scrum and Agile practices, emphasizing the delegation of decision-making authority to the team members who are directly involved in the work. This paradigm shift from traditional hierarchical structures to a more decentralized approach aims to enhance productivity, innovation, and adaptability.

Self-organizing teams operate on the principle that individuals closest to the work possess the most relevant knowledge and are thus best positioned to make informed decisions. This decentralization fosters a sense of ownership and accountability among team members, promoting a collaborative environment where diverse perspectives can converge to solve complex problems.

In the context of Scrum, self-organizing teams are empowered to determine how they will accomplish their tasks within the framework of the Sprint. The Scrum Master facilitates this process by providing guidance and removing impediments, but does not dictate the methods or solutions. This autonomy enables teams to experiment with different approaches, learn from their experiences, and continuously improve their processes.

Empirical research supports the efficacy of self-organizing teams in Agile environments. Studies have demonstrated that teams with greater autonomy exhibit higher levels of motivation, engagement, and job satisfaction. Moreover, these teams tend to deliver higher quality products and are more responsive to changes in customer requirements and market conditions. The iterative nature of Scrum, with its emphasis on regular feedback and incremental progress, aligns well with the dynamic capabilities of self-organizing teams.

However, the transition to self-organizing teams is not without challenges. Organizations accustomed to traditional management styles may encounter resistance to change, as both managers and team members must adjust to new roles and responsibilities. Effective communication and a clear understanding of the principles underlying self-organization are critical to overcoming these hurdles. Training programs, workshops, and coaching can facilitate this transition by providing the necessary skills and knowledge.

Leadership in a self-organizing team context shifts from command-and-control to servant leadership. The role of the Scrum Master exemplifies this shift, as they focus on supporting the team rather than directing it. This involves fostering a culture of trust, encouraging open communication, and facilitating collaboration. Leaders must also be adept at identifying and mitigating obstacles that hinder the team's progress, ensuring that the team can operate at its full potential.

The dynamics of self-organizing teams also necessitate a re-evaluation of performance metrics. Traditional metrics that emphasize individual performance may not be suitable for assessing the effectiveness of self-organizing teams. Instead, metrics that capture team-level outcomes and collaborative efforts are more appropriate. These might include measures of team cohesion, collective problem-solving ability, and the quality of deliverables.

Self-organizing teams thrive in environments that support experimentation and learning. Organizational policies and practices should reflect this by encouraging risk-taking and recognizing the value of learning from failures. Psychological safety is a crucial factor, as it ensures that team members feel comfortable sharing ideas and admitting mistakes without fear of retribution.

The integration of self-organizing teams within the broader Agile framework, particularly the Agile Planning Onion, underscores the importance of alignment between strategic objectives and team-level activities. Each layer of the Planning Onion, from vision and roadmap to iteration and daily planning, requires input and collaboration from self-organizing teams to ensure coherence and alignment with organizational goals.

In conclusion, self-organizing teams represent a fundamental shift in how work is approached and executed in Agile environments. Their ability to adapt, innovate, and deliver high-quality results makes them indispensable to the successful implementation of Scrum and Agile methodologies.

Cross-Functional Skills

Scrum and Agile methodologies underscore the importance of cross-functional skills within development teams. These skills are critical in facilitating the smooth execution of projects and ensuring that teams can adapt to the dynamic nature of software development. Cross-functional skills refer to the ability of team members to perform multiple roles and tasks that may fall outside their primary areas of expertise. This versatility is essential for fostering collaboration and enhancing the overall efficiency of the Scrum team.

In traditional project management, roles and responsibilities are often rigidly defined. However, Agile methodologies advocate for a more fluid approach, where team members are encouraged to develop a broad skill set. This approach not only enables individuals to step in and support various aspects of the project but also fosters a deeper understanding of the project as a whole. For instance, a developer with knowledge of testing can provide valuable insights during the development process, potentially identifying issues early and suggesting improvements. Similarly, a tester with coding skills can contribute to automated testing, accelerating the feedback loop.

Cross-functional skills also play a pivotal role in mitigating risks associated with dependencies. In a Scrum team, tasks are often interdependent, and delays in one area can impact the entire project. By equipping team members with diverse skills, the team becomes more resilient to such disruptions. If a particular task faces a bottleneck due to limited expertise, other team members can step in to alleviate the pressure, ensuring that the project remains on track. This adaptability is crucial in maintaining the momentum of the project and adhering to sprint goals.

Moreover, the cultivation of cross-functional skills fosters a culture of continuous learning and improvement. Scrum teams are encouraged to engage in regular reflection and retrospection, identifying areas for enhancement. By promoting cross-training and skill development, teams can address skill gaps and evolve their capabilities over time. This iterative process aligns with the core principles of Agile, emphasizing continuous improvement and adaptability.

The development of cross-functional skills also enhances team cohesion and collaboration. When team members possess a shared understanding of each other's roles and responsibilities, communication becomes more effective. This shared knowledge base enables more informed decision-making and facilitates the resolution of conflicts. Additionally, it promotes a sense of collective ownership and accountability, as team members are more invested in the success of the project as a whole.

Implementing cross-functional skill development within a Scrum team requires a strategic approach. It involves identifying the key skills needed for the project and providing opportunities for team members to acquire and refine these skills. This can be achieved through various means, such as pair programming, cross-training sessions, and collaborative problem-solving activities. Additionally, organizations can support this development by providing access to relevant training resources and encouraging a culture of knowledge sharing.

In essence, cross-functional skills are integral to the successful implementation of Scrum and Agile methodologies. They enable teams to be more adaptive, resilient, and collaborative, ultimately leading to more efficient and effective project execution. By fostering a culture of continuous learning and skill development, organizations can enhance their Agile practices and achieve sustained success in their software development endeavors.

Collaboration and Communication

Effective collaboration and communication are cornerstones in the successful implementation of Scrum within the Agile Planning Onion framework. The interrelation of these elements fosters a cohesive environment where team members can efficiently work towards shared objectives. This subchapter delves into the mechanisms that enhance collaborative efforts and optimize communication channels, ensuring the seamless execution of Scrum practices.

Scrum inherently promotes teamwork through its roles, events, and artifacts. The Scrum Master, Product Owner, and Development Team must consistently engage in transparent communication to synchronize their efforts. Regular events such as Daily Stand-ups, Sprint Planning, Sprint Reviews, and Retrospectives serve as structured opportunities for dialogue. These interactions are pivotal in aligning team members on project goals, progress, and potential impediments.

The Daily Stand-up, a time-boxed event, enables the team to provide concise updates on their progress and identify any obstacles that may hinder their work. This daily touchpoint ensures that everyone remains informed and can promptly address issues, thereby maintaining the momentum of the sprint. By fostering an environment of open communication, the team can collaboratively problem-solve and adjust their strategies in real-time.

Sprint Planning is another critical event where collaboration is essential. During this session, the Product Owner presents the prioritized backlog items, and the Development Team discusses and estimates the work required to achieve the sprint goal. This collaborative effort ensures that the team has a shared understanding of the tasks at hand and can commit to a realistic sprint backlog. Effective communication during this event is crucial for setting clear expectations and aligning on deliverables.

Sprint Reviews provide a forum for the team to demonstrate the completed work to stakeholders. This event encourages feedback and fosters a collaborative atmosphere between the development team and external parties. Stakeholder input is invaluable for refining the product and aligning it with user needs and business objectives. Transparent communication during Sprint Reviews ensures that the team can make informed decisions about future iterations and adjustments.

Retrospectives offer an opportunity for the team to reflect on the sprint's process and outcomes. This event is integral to continuous improvement, as it allows the team to identify successes and areas for enhancement. Open and honest communication during retrospectives enables the team to collaboratively develop action plans for addressing challenges and optimizing performance in subsequent sprints.

In addition to these structured events, informal communication plays a significant role in fostering collaboration. Encouraging an open-door policy and leveraging communication tools such as instant messaging, video conferencing, and collaborative document platforms can facilitate real-time interactions and information sharing. These tools are particularly beneficial for distributed teams, enabling them to maintain effective communication despite geographical barriers.

The role of the Scrum Master is pivotal in nurturing a collaborative environment. By facilitating events, removing impediments, and promoting a culture of open communication, the Scrum Master ensures that the team can work cohesively and efficiently. This role also involves coaching team members on effective communication practices and fostering an atmosphere of mutual respect and trust.

In the broader context of the Agile Planning Onion, collaboration and communication extend beyond the immediate Scrum team. Engaging with stakeholders, other teams, and organizational leadership is essential for aligning on strategic goals and ensuring that the product development aligns with the overarching vision. Transparent and consistent communication across these layers of the Agile Planning Onion ensures that all parties are informed and can contribute to the project's success.

Effective collaboration and communication are thus fundamental to the successful application of Scrum within the Agile Planning Onion framework. By fostering an environment of openness, transparency, and mutual respect, teams can enhance their collective performance and deliver high-quality products that meet stakeholder expectations.

Delivering Value

The Agile Planning Onion framework is an essential tool for understanding and implementing Scrum methodologies effectively. Within this framework, the concept of delivering value is paramount. Value delivery in Scrum is not merely about completing tasks; it is about ensuring that each increment of work contributes meaningfully to the overarching goals of the project and the organization.

Central to delivering value is the iterative and incremental nature of Scrum. This approach allows for continuous feedback and adaptation, ensuring that the product evolves in alignment with stakeholder needs. Each Sprint, typically lasting two to four weeks, culminates in a potentially shippable product increment. This increment is not just a collection of completed tasks but a coherent enhancement that provides tangible benefits to the end-users and stakeholders.

The Product Owner plays a critical role in value delivery. By maintaining and prioritizing the Product Backlog, the Product Owner ensures that the most valuable items are addressed first. This prioritization is guided by the principle of maximizing return on investment (ROI). The Product Backlog is a dynamic entity, continuously refined based on feedback from stakeholders, changes in market conditions, and technological advancements. This constant refinement ensures that the team is always working on the most valuable features.

Value delivery is also closely tied to the concept of the Definition of Done (DoD). The DoD is a shared understanding within the Scrum Team of what it means for work to be complete. This includes criteria such as code being fully tested, documentation being updated, and the product being deployable. Adhering to the DoD ensures that each increment is of high quality and ready to deliver value upon release. This minimizes technical debt and rework, both of which can detract from the overall value delivered.

Transparency, inspection, and adaptation are the three pillars that support Scrum and are essential for effective value delivery. Transparency ensures that all aspects of the process and the product are visible to those responsible for the outcome. This visibility allows for regular inspection of both the product and the process, enabling timely identification of issues and opportunities for improvement. Adaptation involves making necessary adjustments based on the insights gained from these inspections. Together, these pillars facilitate a continuous cycle of improvement, enhancing the value delivered with each iteration.

Collaboration among team members and with stakeholders is another critical factor. Scrum ceremonies such as Sprint Planning, Daily Stand-ups, Sprint Reviews, and Retrospectives provide structured opportunities for collaboration. During Sprint Planning, the team collectively decides what work will be done in the upcoming Sprint, ensuring alignment with the Product Owner's priorities. Daily Stand-ups allow for quick synchronization and problem-solving, keeping the team on track. Sprint Reviews involve stakeholders in evaluating the product increment, providing valuable feedback that informs future work. Retrospectives focus on process improvement, enabling the team to learn from their experiences and enhance their effectiveness.

Metrics and KPIs also play a vital role in assessing and enhancing value delivery. Common metrics include velocity, burn-down charts, and customer satisfaction scores. These metrics provide quantitative insights into the team's performance and the value being delivered. However, it is essential to use these metrics judiciously, ensuring they drive the right behaviors and support the ultimate goal of delivering value.

In the context of the Agile Planning Onion, value delivery is the culmination of effective planning, prioritization, execution, and continuous improvement. By focusing on delivering value, Scrum teams not only meet project objectives but also contribute to the long-term success and competitiveness of their organizations.

Chapter 10: Tools and Techniques

Agile Project Management Tools

Agile methodologies have revolutionized project management by emphasizing flexibility, collaboration, and customer-centric development. Central to the successful implementation of these methodologies are various tools designed to facilitate the agile process. These tools enable teams to efficiently manage workflows, enhance communication, and ensure the continuous delivery of value.

One of the most prominent tools in the agile ecosystem is Jira. Developed by Atlassian, Jira supports a wide range of agile frameworks, including Scrum and Kanban. It offers robust features for backlog management, sprint planning, and tracking progress through customizable dashboards and reports. Teams can create user stories, tasks, and bugs, assigning them to specific sprints and monitoring their status in real time. The integration capabilities of Jira with other tools such as Confluence, Bitbucket, and various CI/CD pipelines further enhance its utility in fostering a cohesive development environment.

Another vital tool is Trello, which provides a more visual approach to project management through its card-based interface. Trello's boards, lists, and cards allow teams to organize tasks and workflows intuitively. The use of labels, due dates, and checklists within cards helps in tracking progress and ensuring that no detail is overlooked. Trello's simplicity and flexibility make it particularly appealing for smaller teams or projects that require less complex management structures. The tool's integration with other platforms, such as Slack and Google Drive, enhances its functionality and ensures seamless information flow.

Asana is another significant tool that supports agile project management by offering features such as task assignments, project timelines, and workload management. Asana's ability to create dependencies between tasks and visualize project timelines through Gantt charts aids in identifying potential bottlenecks and ensuring timely delivery. The platform's strong emphasis on collaboration is evident through its team communication features and the ability to comment directly on tasks. Asana's integration with tools like Zoom, Microsoft Teams, and various cloud storage services further augments its collaborative capabilities.

For teams that prefer a more Kanban-oriented approach, tools like LeanKit offer extensive support. LeanKit provides visual

Kanban boards that help teams manage work-in-progress limits, identify workflow inefficiencies, and implement continuous improvement practices. The tool's advanced analytics and reporting features enable teams to gain insights into their processes and make data-driven decisions. LeanKit's integrations with tools like Jira and Azure DevOps ensure that it can fit seamlessly into existing workflows.

Azure DevOps, formerly known as Visual Studio Team Services (VSTS), is a comprehensive suite that supports the entire software development lifecycle. It offers features for version control, agile planning, continuous integration, and delivery. Azure Boards, a component of Azure DevOps, provides robust agile planning tools that support Scrum and Kanban methodologies. Teams can manage backlogs, plan sprints, and track progress through customizable boards and dashboards. The integration of Azure Pipelines for CI/CD further enhances the tool's capabilities, ensuring that code changes are continuously tested and deployed.

In addition to these tools, there are numerous other platforms and applications that cater to specific aspects of agile project management. For instance, tools like Slack and Microsoft Teams facilitate real-time communication and collaboration, which are crucial for maintaining alignment and transparency within agile teams. Similarly, tools like Miro and MURAL offer digital whiteboarding capabilities that support remote brainstorming and planning sessions.

The selection of appropriate agile project management tools is contingent on the specific needs and context of the team and project. Regardless of the chosen tools, the overarching goal remains the same: to enhance agility, foster collaboration, and ensure the continuous delivery of value to customers.

Collaboration Tools

In the realm of Agile methodologies, efficient collaboration stands as a cornerstone for successful project execution. Scrum, a widely adopted Agile framework, emphasizes the necessity for seamless communication and coordination among team members. Consequently, the implementation of collaboration tools becomes indispensable. These tools facilitate real-time interaction, transparency, and synchronization, which are critical for the iterative and incremental nature of Scrum.

One of the primary tools utilized in Scrum is the task board, often digital, that visualizes the workflow. This board typically comprises columns representing different stages of the development process, such as "To Do," "In Progress," and "Done." Teams frequently employ software solutions like Jira, Trello, or Azure DevOps to maintain these boards. These platforms not only provide a visual representation of task progression but also allow for easy updating and tracking of individual responsibilities and deadlines. The visibility afforded by task boards ensures that all team members, including stakeholders, are aware of the current status of each task, thereby fostering accountability and transparency.

Another vital tool in the Scrum toolkit is the backlog management system. The product backlog and sprint backlog are essential artifacts in Scrum, and tools like ProductPlan or Pivotal Tracker are often used to manage them effectively. These tools enable the product owner to prioritize tasks based on value and urgency, ensuring that the team focuses on the most critical items first. Additionally, they provide functionalities for estimating effort, setting priorities, and tracking progress, which are crucial for maintaining the momentum of the project.

Communication platforms such as Slack, Microsoft Teams, or Zoom play a pivotal role in facilitating daily stand-ups, sprint planning, and retrospective meetings. These tools support synchronous and asynchronous communication, allowing team members to stay connected regardless of their geographical locations. Video conferencing features are particularly beneficial for conducting face-to-face meetings, which are a core component of Scrum ceremonies. Screen sharing, file sharing, and integrated chat functionalities further enhance the collaborative experience, enabling teams to resolve issues promptly and make informed decisions.

Documentation and knowledge-sharing tools like Confluence or SharePoint are also integral to Scrum. These platforms serve as repositories for storing project-related documents, meeting notes, and other critical information. They support version control and collaborative editing, ensuring that the documentation is always up-to-date and accessible to all team members. This centralized knowledge base is invaluable for onboarding new team members, maintaining project continuity, and preserving institutional knowledge.

Additionally, time-tracking and reporting tools such as Toggl or Harvest are employed to monitor the time spent on various tasks and activities. These tools provide insights into team productivity and help in identifying bottlenecks or inefficiencies in the workflow. Accurate time tracking is essential for evaluating the team's velocity and for making data-driven adjustments to the sprint planning process.

The integration of these tools within a cohesive ecosystem is paramount. Most modern collaboration tools offer APIs and integration capabilities, allowing them to work seamlessly together. For instance, task boards can be integrated with communication platforms to provide real-time notifications about task updates, or time-tracking tools can be linked with backlog management systems to correlate time spent with task completion. This interconnected environment enhances the overall efficiency of the Scrum process, enabling teams to deliver high-quality products in a timely manner.

In essence, the judicious selection and implementation of collaboration tools are critical for the success of Scrum. These tools not only facilitate effective communication and coordination but also provide the necessary infrastructure for managing tasks, documentation, and time. By leveraging these tools, Scrum teams can achieve a high level of transparency, accountability, and productivity, ultimately leading to the successful delivery of projects.

Metrics and Reporting Tools

The effective implementation of Scrum and the Agile Planning Onion necessitates the utilization of robust metrics and reporting tools. These instruments are indispensable for tracking progress, identifying bottlenecks, and ensuring alignment with project goals. Metrics and reporting tools offer quantitative insights that facilitate data-driven decision-making, which is crucial for maintaining the agility and responsiveness that underpin Scrum methodologies.

Central to the measurement of Scrum effectiveness are key performance indicators (KPIs) such as velocity, sprint burndown, and cumulative flow diagrams. Velocity measures the amount of work a team can complete in a single sprint, providing a benchmark for future planning and forecasting. It is calculated by summing the story points of all completed user stories at the end of the sprint. While velocity offers valuable insights, it should be interpreted with caution, as it can be influenced by factors such as team composition, complexity of tasks, and external dependencies.

Sprint burndown charts visually represent the progress of a sprint by plotting remaining work against time. This tool enables teams to monitor their pace and detect any deviations from the planned trajectory. A well-maintained burndown chart can highlight potential impediments early in the sprint, allowing for timely interventions. To augment the utility of burndown charts, teams often employ daily stand-up meetings to discuss progress and adjust plans as necessary.

Cumulative flow diagrams (CFDs) provide a comprehensive overview of work in progress (WIP) across different stages of the workflow. By illustrating the rate at which tasks move through the system, CFDs help teams identify bottlenecks and optimize their processes. The area between the curves in a CFD represents the WIP, and a steady flow indicates a balanced system. Conversely, an increasing WIP suggests potential issues that require attention.

In addition to these primary metrics, secondary metrics such as lead time, cycle time, and defect density offer further granularity. Lead time measures the duration from the inception of a task to its completion, while cycle time focuses on the period from the start of active work to its conclusion. Both metrics are pivotal for understanding and improving efficiency. Defect density, calculated as the number of defects per unit of work, provides insights into the quality of deliverables and the effectiveness of testing practices.

The integration of reporting tools with these metrics enhances their accessibility and interpretability. Tools such as Jira, Trello, and Azure DevOps offer built-in features for tracking and visualizing metrics. These platforms facilitate real-time updates and collaboration, ensuring that all team members have a shared understanding of progress and challenges. Custom dashboards and reports can be tailored to meet the specific needs of the team, providing actionable insights at a glance.

Automated reporting tools further streamline the process of data collection and analysis. By minimizing manual input, automation reduces the risk of errors and frees up time for more value-adding activities. Advanced analytics and machine learning capabilities embedded in some tools can predict trends and suggest optimizations, thereby enhancing the team's ability to respond proactively to emerging issues.

The adoption of metrics and reporting tools is not without challenges. Teams must select appropriate metrics that align with their goals and context. Over-reliance on a single metric can lead to skewed priorities and unintended consequences. It is essential to maintain a balanced perspective, considering both quantitative and qualitative data. Regular retrospectives provide an opportunity to reflect on the effectiveness of the chosen metrics and adjust them as necessary.

Effective metrics and reporting tools are integral to the successful application of Scrum and the Agile Planning Onion. They provide the empirical foundation needed to navigate the complexities of agile project management, enabling teams to deliver high-quality outcomes in a timely and efficient manner.

Automation in Agile

Automation has become an integral component in the Agile methodology, contributing significantly to the efficiency and effectiveness of Scrum practices. This integration aligns with the core principles of Agile, which emphasize iterative development, continuous feedback, and rapid delivery of functional software.

Automation in Agile encompasses various aspects, including test automation, continuous integration/continuous deployment (CI/CD), and automated documentation, each playing a pivotal role in enhancing the overall workflow.

Test automation serves as a cornerstone in Agile environments, ensuring that software is consistently tested throughout the development lifecycle. Automated tests, ranging from unit tests to integration and system tests, facilitate early detection of defects, thereby reducing the risk of regression and ensuring that new changes do not adversely affect existing functionality. By leveraging test automation, teams can execute a comprehensive suite of tests rapidly and frequently, supporting the Agile principle of delivering incremental value in short cycles.

Continuous integration and continuous deployment (CI/CD) pipelines further streamline the Agile process by automating the integration and deployment phases. CI involves the frequent merging of code changes into a shared repository, followed by automated builds and tests. This practice ensures that integration issues are identified and resolved early, maintaining a stable codebase. CD extends this automation to the deployment phase, enabling automatic delivery of code changes to production or staging environments. This continuous delivery mechanism aligns with Agile's goal of maintaining a deployable product at all times, facilitating rapid feedback from end-users and stakeholders.

Automated documentation tools also play a crucial role in Agile practices. These tools generate up-to-date documentation based on the current state of the codebase, ensuring that documentation remains consistent with the software's actual implementation. Automated documentation reduces the manual effort involved in maintaining accurate and timely documentation, allowing teams to focus more on development activities. This approach supports Agile's emphasis on working software over comprehensive documentation, while still providing necessary information for future reference and knowledge transfer.

The integration of automation in Agile practices necessitates a cultural shift within development teams. It requires a commitment to continuous improvement, a willingness to adopt new tools and technologies, and a focus on collaboration and communication. Teams must invest in training and skill development to effectively implement and maintain automated processes. This cultural shift is underpinned by the Agile principle of fostering a collaborative and empowered team environment, where members are encouraged to experiment, learn, and adapt.

Moreover, the adoption of automation in Agile is not without challenges. Teams may encounter obstacles such as tool compatibility issues, integration complexities, and the initial investment in setting up automated systems. However, these challenges can be mitigated through careful planning, incremental implementation, and continuous monitoring and improvement. By addressing these challenges proactively, teams can reap the long-term benefits of automation, including increased efficiency, higher-quality software, and faster delivery cycles.

In conclusion, automation is a critical enabler of Agile methodologies, enhancing the efficiency, quality, and speed of software development. By integrating test automation, CI/CD pipelines, and automated documentation, Agile teams can achieve a more streamlined and effective workflow. This integration requires a cultural shift and a proactive approach to overcoming challenges, but the benefits of automation in Agile are substantial and far-reaching.

Chapter 11: Scaling Agile and Scrum

Frameworks for Scaling Agile

Agile methodologies, particularly Scrum, have become integral to modern software development practices. As organizations seek to leverage these methodologies across larger and more complex projects, the necessity for frameworks that support the scaling of Agile practices becomes paramount. This subchapter explores several prominent frameworks designed to facilitate the application of Agile principles in large-scale environments.

One of the most widely recognized frameworks for scaling Agile is the Scaled Agile Framework (SAFe). SAFe provides a structured approach for scaling Agile across multiple teams, departments, and even entire enterprises. It incorporates principles from Lean, Agile, and product development flow to support alignment, collaboration, and delivery across a large number of Agile teams. SAFe is built around four core values: alignment, built-in quality, transparency, and program execution. These values are operationalized through a series of levels—Team, Program, Large Solution, and Portfolio—each with specific roles, responsibilities, and practices to ensure coherent and efficient scaling.

Another prominent scaling framework is Large-Scale Scrum (LeSS). LeSS extends the principles of Scrum while maintaining its simplicity. It emphasizes the importance of feature teams, which are cross-functional teams capable of delivering end-to-end customer features. LeSS operates on two levels: the LeSS framework for up to eight teams and LeSS Huge for more extensive implementations. The framework promotes deep customer focus, continuous improvement, and minimizing waste, aligning closely with Lean principles. LeSS encourages minimal management overhead, advocating for a flat organizational structure to enhance communication and collaboration.

Disciplined Agile Delivery (DAD) is another framework that provides a more prescriptive approach to scaling Agile. DAD encompasses a broader range of life cycle choices, including Scrum, Kanban, and Lean, allowing teams to choose the most appropriate method for their context. It emphasizes a goal-driven approach, where the choice of practices and processes is guided by the specific goals and context of the project. DAD also incorporates governance and risk management practices, ensuring that scaled Agile implementations remain compliant and aligned with organizational objectives.

Nexus, developed by Scrum.org, is a framework designed to scale Scrum itself. Nexus focuses on the integration of work produced by multiple Scrum teams working on a single product. It introduces the Nexus Integration Team, responsible for ensuring that the integrated increment is delivered each Sprint. This team also addresses dependencies and integration issues, facilitating seamless collaboration among Scrum teams. Nexus retains the simplicity and core principles of Scrum while providing additional structure for managing inter-team dependencies and integration challenges.

Each of these frameworks offers unique approaches and tools for scaling Agile practices. The choice of framework depends on various factors, including the size of the organization, the complexity of the projects, and the existing organizational culture. While SAFe provides a comprehensive and structured approach, LeSS focuses on simplicity and minimizing waste. DAD offers flexibility in choosing life cycles and practices, and Nexus emphasizes integration and dependency management.

Understanding these frameworks allows organizations to make informed decisions about how to scale Agile practices effectively. The successful implementation of any of these frameworks requires a deep understanding of Agile principles, a commitment to continuous improvement, and the flexibility to adapt practices to the specific needs and context of the organization.

Challenges in Scaling

Scaling Scrum and Agile methodologies to larger projects or entire organizations presents several significant challenges. One primary issue is maintaining the core principles of Agile, such as flexibility, collaboration, and rapid iteration, when the number of teams and stakeholders increases. This often requires a shift from small, closely-knit teams to a more distributed and complex structure, which can dilute the effectiveness of Agile practices if not managed correctly.

Coordination across multiple teams becomes a critical factor. In smaller projects, teams can operate semi-autonomously with minimal overlap. However, as the scale increases, dependencies between teams grow, necessitating more structured communication and synchronization mechanisms. This can lead to the need for additional roles or frameworks, such as Scrum of Scrums or the Scaled Agile Framework (SAFe), to ensure alignment and coherence across the entire project.

Another challenge is the consistency of the Agile mindset across all levels of the organization. While individual teams may adopt Agile practices, ensuring that middle and upper management also understand and support these principles is crucial. Often, traditional management practices and mindsets can conflict with Agile values, leading to resistance or half-hearted implementation. This necessitates a cultural shift, which can be difficult to achieve and sustain without strong leadership and continuous education.

Resource allocation and prioritization become more complex at scale. In smaller teams, prioritization can be managed through direct communication and consensus. In larger settings, however, aligning priorities across multiple teams and ensuring that resources are allocated effectively requires more sophisticated planning and governance structures. This can sometimes result in a return to more hierarchical decision-making processes, which can be at odds with Agile's decentralized nature.

Technical debt and integration issues are amplified in larger projects. As more teams contribute to the codebase, the risk of inconsistencies and integration problems increases. Continuous integration and continuous deployment (CI/CD) practices become essential, but their implementation can be challenging. Ensuring that all teams adhere to the same standards and practices requires rigorous oversight and often, additional tools and infrastructure.

Metrics and performance evaluation also pose a challenge. While Agile emphasizes qualitative measures such as customer satisfaction and team velocity, scaling these metrics to a larger context can be difficult. Organizations may be tempted to revert to traditional quantitative metrics, which can undermine the Agile approach. Developing appropriate metrics that reflect the health and progress of Agile at scale is essential but complex.

Finally, training and skill development must be addressed. Scaling Agile involves not just increasing the number of teams but also ensuring that all team members possess the necessary skills and knowledge. This can require significant investment in training programs and continuous learning opportunities. Additionally, fostering a community of practice within the organization can help share knowledge and best practices, further supporting the scaling effort.

In essence, scaling Scrum and Agile methodologies involves navigating a series of interrelated challenges that require careful planning, strong leadership, and a commitment to the core principles of Agile. Balancing the need for structure and coordination with the flexibility and responsiveness that Agile promotes is critical to successfully scaling these methodologies across larger projects and organizations.

Best Practices for Scaling

Scaling Scrum and Agile methodologies to larger projects and organizations introduces complexities that smaller teams may not encounter. To address these, several best practices have been identified to ensure that the principles of Scrum and the Agile Planning Onion are effectively maintained across broader contexts.

First, it is essential to maintain the integrity of the Scrum framework. This involves adhering strictly to the roles, events, and artifacts defined in Scrum. When scaling, the temptation to modify or dilute these elements can be strong, but doing so often leads to a breakdown in communication and a loss of the benefits that Scrum provides. For instance, the role of the Scrum Master becomes even more critical in a scaled environment as they ensure that Scrum practices are being followed and facilitate coordination between multiple teams.

Second, the Agile Planning Onion must be applied consistently across all levels of the organization. This means that the strategic, portfolio, product, release, iteration, and daily planning layers must be aligned and integrated. At the strategic level, leadership should provide a clear vision and objectives that guide the portfolio and product planning. Portfolio management must prioritize initiatives based on value and alignment with strategic goals. Product owners should then translate these priorities into actionable product backlogs that guide release and iteration planning. Daily planning ensures that teams stay on track and can adapt to any changes promptly.

Third, the use of scaling frameworks such as SAFe (Scaled Agile Framework), LeSS (Large Scale Scrum), or Nexus can provide structured approaches to scaling Agile practices. These frameworks offer guidelines and practices that help maintain alignment, quality, and productivity across multiple teams. For example, SAFe introduces roles such as Release Train Engineer and Product Management, which help coordinate efforts across teams and ensure that the organization's strategic goals are being met.

Fourth, fostering a culture of continuous improvement and learning is vital. As organizations scale, the complexity of projects and the interdependence of teams increase. Regular retrospectives at the team level should be complemented by broader organizational reviews to identify and address systemic issues. Encouraging a culture where feedback is valued and acted upon helps teams to continuously refine their processes and improve performance.

Fifth, effective communication and collaboration tools are essential in a scaled environment. Tools such as Jira, Confluence, and Slack can facilitate coordination and information sharing across teams. These tools should be used to maintain transparency and ensure that all team members have access to the information they need to perform their roles effectively. Additionally, regular cross-team meetings, such as Scrum of Scrums, can help ensure that dependencies are managed, and teams are aligned.

Sixth, managing dependencies and synchronizing efforts across teams requires careful planning and coordination. Techniques such as dependency mapping and program increment planning can help identify and manage dependencies early in the planning process. Synchronization mechanisms such as cadence and synchronization points ensure that teams are working in harmony and can deliver integrated increments of value.

Lastly, leadership plays a crucial role in scaling Agile practices. Leaders must champion the principles of Scrum and the Agile Planning Onion and provide the necessary support and resources for teams to succeed. This includes investing in training and development, removing impediments, and fostering an environment where Agile practices can thrive.

By adhering to these best practices, organizations can successfully scale Scrum and Agile methodologies, ensuring that they can deliver value effectively and efficiently, even as the complexity and scope of their projects grow.

Case Studies in Scaling Agile

Scaling Agile methodologies, particularly Scrum, within large organizations presents both opportunities and challenges. This subchapter examines empirical case studies to elucidate the practical applications and outcomes of scaling Agile. These studies provide valuable insights into the methodologies employed, the obstacles encountered, and the strategies devised to address these challenges.

One notable case involves a global financial services firm that sought to implement Scrum across multiple departments. Initially, the firm faced resistance due to ingrained traditional project management practices. The firm employed a phased rollout strategy, beginning with pilot teams to demonstrate the efficacy of Scrum. These pilot teams served as internal champions, showcasing early successes in terms of increased productivity and improved stakeholder satisfaction. The firm subsequently scaled Scrum by forming a dedicated Agile transformation office to provide ongoing training and support. This approach facilitated a smoother transition and fostered a culture of continuous improvement.

Another case study focuses on a multinational technology company that aimed to integrate Agile practices across its diverse product lines. The company adopted the Scaled Agile

Framework (SAFe) to synchronize the efforts of multiple Scrum teams working on interdependent projects. The implementation of SAFe involved rigorous training sessions for all team members and the establishment of a Program Increment (PI) planning cadence. This structure enabled better alignment of business objectives with development activities, resulting in accelerated product delivery cycles and enhanced cross-team collaboration.

A third case study examines a healthcare organization that implemented Scrum to streamline its software development processes. The organization encountered significant challenges related to regulatory compliance and stringent documentation requirements. To address these issues, the Scrum teams incorporated compliance checkpoints into their Definition of Done (DoD) criteria. This adaptation ensured that all regulatory requirements were met without compromising the Agile principles of iterative development and customer feedback. The organization also leveraged automated testing tools to maintain high-quality standards, thereby reducing the time and effort required for manual compliance verification.

An additional case involves a telecommunications company that aimed to scale Agile practices to improve its network infrastructure projects. The company adopted the Large-Scale Scrum (LeSS) framework, which emphasizes simplicity and the minimization of dependencies between teams. The implementation of LeSS required a reorganization of teams into feature-based groups, each responsible for delivering end-to-end functionality. This reorganization was complemented by frequent cross-team synchronization meetings to address any interdependencies and ensure cohesive progress. The adoption of LeSS resulted in faster delivery times and improved adaptability to changing market demands.

These case studies underscore the importance of context-specific adaptations when scaling Agile methodologies. While frameworks like SAFe and LeSS provide structured approaches, the unique characteristics of each organization necessitate tailored strategies. Common themes across these cases include the critical role of leadership support, the necessity of continuous training and coaching, and the importance of fostering a collaborative culture. Additionally, the integration of compliance and quality assurance measures into the Agile workflow is crucial for organizations operating in highly regulated industries.

The analysis of these case studies highlights that successful scaling of Agile requires a multifaceted approach. Organizations must be prepared to experiment, learn from their experiences, and iteratively refine their strategies. By leveraging the insights derived from these empirical examples, organizations can enhance their ability to scale Agile practices effectively, thereby achieving greater agility, improved product quality, and heightened customer satisfaction.

Chapter 12: Common Pitfalls and How to Avoid Them

Overcoming Resistance to Change

Resistance to change is a common challenge encountered during the implementation of Scrum and Agile methodologies. Understanding the underlying causes and strategies to mitigate this resistance is crucial for successful adoption. Organizations often face resistance due to a variety of factors, including fear of the unknown, loss of control, and entrenched habits. Addressing these concerns systematically can facilitate a smoother transition to Agile practices.

One of the primary reasons for resistance is fear of the unknown. Employees accustomed to traditional project management methodologies may feel uncertain about how Agile practices will impact their roles and responsibilities. Providing comprehensive training and education about Scrum and Agile principles can alleviate these fears. Workshops, seminars, and hands-on training sessions can help demystify Agile processes, making them more approachable and less intimidating.

Loss of control is another significant factor contributing to resistance. Traditional management structures often involve a top-down approach, where decisions are made by a few individuals at the top of the hierarchy. Agile methodologies, on the other hand, promote a more decentralized decision-making process. This shift can be perceived as a loss of control by managers and supervisors. To mitigate this, it is essential to communicate the benefits of decentralization, such as increased team autonomy, faster decision-making, and enhanced innovation. Engaging managers in the transition process and involving them in Agile training can also help them understand and embrace their evolving roles.

Entrenched habits and established workflows pose another challenge. Employees may be resistant to changing long-standing practices and routines. Incremental implementation of Agile practices can be an effective strategy to overcome this resistance. Instead of a complete overhaul, introducing Agile elements gradually allows employees to adapt at a manageable pace. For instance, starting with daily stand-up meetings or incorporating iterative planning sessions can provide a gentle introduction to Agile methodologies.

Effective communication is paramount in overcoming resistance to change. Transparent and consistent communication about the reasons for adopting Agile, the expected benefits, and the anticipated challenges can help build trust and buy-in from the team. Regular updates and open forums for discussion can address concerns and provide a platform for employees to voice their opinions and ask questions.

Leadership plays a critical role in facilitating change. Leaders who actively champion Agile practices and demonstrate commitment to the transition can inspire confidence in their teams. Modeling Agile behaviors, such as collaboration, flexibility, and continuous improvement, sets a positive example and reinforces the importance of the change.

In addition to leadership support, involving employees in the change process can enhance their sense of ownership and reduce resistance. Empowering teams to contribute to the design and implementation of Agile practices can foster a collaborative environment and increase engagement. Encouraging feedback and incorporating employee suggestions into the transition plan can further reinforce their commitment to the change.

Addressing resistance also involves recognizing and rewarding early adopters and successes. Celebrating small wins and acknowledging the efforts of individuals and teams who embrace Agile practices can create a positive momentum. This recognition can motivate others to follow suit and reduce resistance over time.

In conclusion, overcoming resistance to change requires a multifaceted approach that addresses the fears, concerns, and habits of employees. Comprehensive training, effective communication, leadership support, incremental implementation, and employee involvement are key strategies to facilitate a successful transition to Agile methodologies. By addressing these factors, organizations can create an environment conducive to the adoption of Scrum and Agile practices, ultimately leading to improved project outcomes and organizational performance.

Avoiding Scope Creep

Scope creep represents one of the most insidious challenges in project management, particularly within Agile frameworks such as Scrum. It refers to the uncontrolled expansion of project scope without corresponding adjustments to time, cost, and resources. This phenomenon can undermine the integrity of the Agile Planning Onion, a conceptual model that delineates the layers of planning in Agile projects. Effective strategies to mitigate scope creep are therefore essential for maintaining project coherence and achieving desired outcomes.

Central to avoiding scope creep in Scrum is the rigorous application of time-boxed iterations, known as sprints. Each sprint should be clearly defined in terms of deliverables, with a fixed duration and a commitment to not alter the sprint goal once it has commenced. This temporal constraint enforces discipline and ensures that the team remains focused on a specific subset of features or tasks, thereby limiting the potential for scope expansion within the sprint period.

Another critical element is the role of the Product Owner (PO). The PO is tasked with maintaining and prioritizing the Product Backlog, ensuring that only the most valuable and necessary features are included in the project scope. By continually refining and prioritizing the backlog items, the PO can effectively manage stakeholder expectations and prevent the inclusion of extraneous requirements. Regular backlog grooming sessions, held in collaboration with the development team, provide an opportunity to reassess and adjust priorities based on current project realities and stakeholder feedback.

Moreover, the Definition of Done (DoD) serves as a safeguard against scope creep. A well-articulated DoD sets clear criteria for what constitutes a completed task, feature, or user story. By adhering strictly to these criteria, the team can avoid the temptation to add 'just one more feature' or make last-minute changes that could jeopardize the sprint goal. The DoD should be a living document, revisited and refined as the team gains more experience and insight into the project.

Effective communication is also paramount in combating scope creep. Regular and structured communication channels, such as daily stand-ups, sprint reviews, and retrospectives, foster a transparent environment where potential deviations from the project scope can be identified and addressed promptly. These meetings provide a forum for the team to discuss progress, identify obstacles, and realign their efforts with the sprint and product goals.

Change management processes are another essential tool in the Scrum arsenal. Any proposed changes to the project scope should undergo a formal evaluation process, where the potential impact on time, cost, and resources is assessed. This structured approach ensures that changes are not made impulsively and that all stakeholders understand the implications of any scope adjustments.

Lastly, fostering a culture of accountability and ownership within the team can significantly reduce the risk of scope creep. When team members take responsibility for their work and understand the importance of adhering to the defined scope, they are more likely to resist pressures to expand the project boundaries. This cultural shift can be reinforced through continuous education and training on Agile principles and best practices.

In essence, avoiding scope creep in Scrum requires a multifaceted approach that combines disciplined planning, effective communication, and a strong adherence to Agile principles. By implementing these strategies, teams can maintain control over their project scope, thus ensuring the successful delivery of valuable and high-quality products.

Ensuring Team Alignment

Ensuring that all team members are aligned is a critical factor in the successful implementation of Scrum and the Agile Planning Onion. Alignment within a team ensures that everyone shares a common understanding of the objectives, processes, and expectations, thereby fostering a collaborative and efficient working environment. This subchapter delves into the mechanisms and strategies that can be employed to achieve and maintain this alignment.

One of the primary tools for ensuring alignment is the Sprint Planning Meeting. This meeting serves as a forum where the team collectively decides on the tasks to be undertaken in the upcoming sprint. During this meeting, the Product Owner presents the prioritized Product Backlog items, and the team collaborates to estimate and select the items they commit to completing. The Sprint Planning Meeting not only clarifies the scope of work but also aligns the team on the objectives and expectations for the sprint.

Another critical element in maintaining alignment is the Daily Scrum, also known as the Daily Stand-up. This short, time-boxed meeting provides a regular checkpoint for team members to synchronize their activities, discuss progress, and identify any impediments. The Daily Scrum helps to ensure that everyone is on the same page and allows for quick adjustments to be made if necessary. It reinforces the shared commitment to the sprint goals and fosters transparency and accountability among team members.

The Sprint Review Meeting is another pivotal event in the Scrum framework that contributes to team alignment. During the Sprint Review, the team demonstrates the work completed during the sprint to the Product Owner and other stakeholders. This meeting provides an opportunity for feedback and ensures that the delivered increment meets the stakeholders' expectations. The Sprint Review fosters a shared understanding of the project's progress and allows for adjustments based on stakeholder input, thereby aligning the team with the broader project goals.

The Sprint Retrospective is an essential practice for maintaining alignment by reflecting on the team's processes and performance. In this meeting, the team discusses what went well, what did not go well, and what can be improved in future sprints. The Retrospective encourages continuous improvement and helps the team to align on best practices and process enhancements. By addressing issues and identifying opportunities for improvement, the team can better align their efforts with the overall project objectives.

Effective communication is a cornerstone of team alignment. Regular and transparent communication helps to ensure that all team members are aware of the project's status, any changes in priorities, and any challenges that may arise. Tools such as collaborative software, shared documents, and communication platforms can facilitate this transparency. Ensuring that information flows freely within the team helps to prevent misunderstandings and keeps everyone aligned with the project goals.

Moreover, the role of the Scrum Master is crucial in fostering and maintaining team alignment. The Scrum Master acts as a facilitator, ensuring that Scrum practices are followed and that any impediments to progress are removed. By promoting a culture of collaboration and continuous improvement, the Scrum Master helps to align the team's efforts with the Agile principles and the project's objectives.

In conclusion, ensuring team alignment is vital for the success of Scrum and the Agile Planning Onion. Through structured meetings, effective communication, and the facilitative role of the Scrum Master, teams can achieve and maintain alignment, thereby enhancing their ability to deliver high-quality products efficiently and effectively.

Maintaining Sustainable Pace

The principle of maintaining a sustainable pace is integral to the effective implementation of Scrum within the Agile framework. This concept, derived from the Agile Manifesto, emphasizes the importance of creating a work environment where teams can maintain a consistent and manageable workload over an extended period. This approach not only enhances productivity but also fosters a healthy work-life balance, reducing the risk of burnout and promoting long-term team cohesion.

Scientific studies have consistently shown that prolonged periods of overwork can lead to significant declines in productivity and creativity. The human cognitive system requires adequate rest to function optimally, and chronic stress can impair decision-making processes and problem-solving abilities. In the context of Scrum, maintaining a sustainable pace ensures that team members remain engaged and motivated, thereby improving the overall quality of deliverables.

One of the key mechanisms for maintaining a sustainable pace in Scrum is the implementation of time-boxed iterations, known as sprints. Sprints, typically lasting two to four weeks, provide a structured timeframe within which specific goals must be achieved. This time-boxing approach helps teams manage their workload more effectively by breaking down large projects into smaller, manageable tasks. It also allows for regular assessment and adjustment, ensuring that the pace remains sustainable.

The role of the Scrum Master is pivotal in monitoring and facilitating a sustainable pace. The Scrum Master must ensure that the team adheres to the principles of Agile and Scrum, including the maintenance of a balanced workload. This involves regular communication with team members to assess their workload and stress levels, as well as the identification and removal of impediments that may hinder progress. The Scrum Master must also foster an environment where team members feel comfortable discussing their capacity and any challenges they face.

Empirical evidence supports the notion that maintaining a sustainable pace leads to higher levels of team satisfaction and performance. A study conducted by the Journal of Occupational Health Psychology found that employees who experienced a balanced workload reported higher job satisfaction and lower levels of stress and burnout. These findings underscore the importance of creating a work environment that prioritizes sustainable practices.

Moreover, maintaining a sustainable pace is closely linked to the concept of continuous improvement, or Kaizen, in Agile methodologies. By regularly reflecting on their processes and outcomes through retrospectives, Scrum teams can identify areas for improvement and implement changes incrementally. This iterative process not only enhances efficiency but also ensures that the pace of work remains manageable over time.

In practical terms, teams can employ several strategies to maintain a sustainable pace. These include setting realistic goals, prioritizing tasks effectively, and ensuring regular breaks and downtime. The use of velocity tracking, which measures the amount of work completed in each sprint, can also provide valuable insights into the team's capacity and help in planning future sprints more accurately.

The alignment of sustainable pace with the core values of Scrum and Agile—such as respect, commitment, and focus—reinforces its importance in the methodology. By prioritizing sustainable practices, Scrum teams can achieve a harmonious balance between productivity and well-being, ultimately leading to more successful and enduring project outcomes. The integration of sustainable pace principles into the Agile Planning

Onion further underscores their relevance and necessity in the pursuit of effective and resilient project management.

Chapter 13: Future Trends in Agile and Scrum

Emerging Practices

Scrum and Agile methodologies have been pivotal in reshaping the landscape of project management and software development. Emerging practices within this domain are continuously evolving, driven by the need for more efficient, adaptable, and collaborative approaches. These practices are often characterized by their ability to respond to complex and rapidly changing project requirements, while maintaining a focus on delivering high-quality outcomes.

One notable emerging practice is the integration of DevOps within Agile frameworks. DevOps, which emphasizes the collaboration between development and operations teams, seeks to streamline the software delivery process. By incorporating DevOps principles into Scrum, teams can achieve continuous integration and continuous delivery (CI/CD), reducing the time between writing code and deploying it to production. This integration fosters a culture of shared responsibility, enhancing both the speed and reliability of software releases.

Another significant development is the adoption of Behavior-Driven Development (BDD). BDD extends Test-Driven Development (TDD) by using natural language constructs to define the behavior of software. This practice encourages collaboration among developers, testers, and business stakeholders, ensuring that the software meets the intended requirements. BDD scenarios serve as living documentation, providing a clear and shared understanding of how the system should behave under various conditions.

The use of Agile metrics and data-driven decision-making has also gained traction. Traditional metrics, such as velocity and burndown charts, are being supplemented with more sophisticated analytics. These include cycle time, lead time, and cumulative flow diagrams, which provide deeper insights into the team's performance and the flow of work. By leveraging these metrics, teams can identify bottlenecks, predict future performance, and make informed adjustments to their processes.

Scaled Agile Frameworks (SAFe) and Large-Scale Scrum (LeSS) are increasingly being adopted by organizations seeking to implement Agile practices across multiple teams and departments. These frameworks provide structured approaches to scaling Agile, ensuring alignment and coherence across the enterprise. They address challenges such as inter-team dependencies, governance, and coordination, facilitating the successful implementation of Agile at scale.

Additionally, the role of Agile coaches and facilitators has become more prominent. These professionals provide guidance and support to teams, helping them navigate the complexities of Agile adoption. They play a crucial role in fostering a culture of continuous improvement, encouraging teams to experiment with new practices and refine their processes. Agile coaches also assist in resolving conflicts, enhancing communication, and promoting a collaborative environment.

The integration of Artificial Intelligence (AI) and Machine Learning (ML) into Agile practices is an emerging trend with significant potential. AI and ML can be used to automate repetitive tasks, analyze large datasets, and provide predictive insights. For example, AI-driven tools can help in identifying patterns in user stories, suggesting optimal sprint planning, and forecasting project risks. These technologies augment the capabilities of Agile teams, enabling them to make data-driven decisions and improve their overall efficiency.

Furthermore, there is a growing emphasis on sustainability and ethical considerations within Agile practices. Teams are increasingly aware of the environmental and social impacts of their work. Sustainable Agile practices involve minimizing waste, optimizing resource usage, and considering the long-term effects of software products. Ethical considerations encompass data privacy, security, and inclusivity, ensuring that the software developed aligns with societal values and principles.

The continuous evolution of Agile practices reflects the dynamic nature of the field. As new challenges and opportunities arise, Agile teams must remain adaptable, leveraging emerging practices to enhance their effectiveness and deliver value. These developments underscore the importance of a flexible, iterative approach to project management, where learning and improvement are integral to success.

Technological Advancements

The evolution of Scrum and the Agile Planning Onion is inextricably linked with technological advancements that have revolutionized the way teams collaborate, communicate, and deliver value. These advancements have facilitated more efficient workflows, enhanced transparency, and improved stakeholder engagement, which are critical components of Agile methodologies.

One of the most significant technological contributions to Scrum is the development of sophisticated project management tools. Platforms such as Jira, Trello, and Asana have become indispensable for Agile teams. These tools provide digital backlogs, sprint planning boards, and real-time tracking of tasks and user stories, enabling teams to maintain a clear and up-to-date view of project progress. The ability to visualize workflows and monitor key performance indicators (KPIs) ensures that teams can adapt quickly to changing circumstances, a core tenet of Agile principles.

Cloud computing has also played a pivotal role in the advancement of Agile practices. By leveraging cloud-based services, teams can access project resources and collaborate from anywhere in the world. This has been particularly beneficial for distributed teams, allowing for seamless integration and communication across different time zones and geographical locations. Cloud infrastructure supports continuous integration and continuous deployment (CI/CD) pipelines, enabling rapid iterations and faster delivery of high-quality software.

Automation tools have further enhanced the efficiency of Scrum processes. Automated testing frameworks, such as Selenium and JUnit, ensure that code is consistently tested and validated, reducing the likelihood of defects and technical debt. Continuous integration servers like Jenkins and GitLab CI automate the build and deployment processes, allowing for frequent and reliable releases. Automation not only accelerates development cycles but also frees up team members to focus on more complex and creative tasks.

Artificial intelligence (AI) and machine learning (ML) technologies are beginning to influence Agile methodologies as well. Predictive analytics can forecast project timelines, identify potential risks, and suggest optimal resource allocation. AI-driven tools can analyze historical project data to provide insights into team performance and areas for improvement. These capabilities enable more informed decision-making and proactive management of the development process.

Communication technologies have also seen significant advancements, contributing to more effective collaboration within Agile teams. Tools like Slack, Microsoft Teams, and Zoom facilitate instant messaging, video conferencing, and file sharing, ensuring that team members can communicate effortlessly regardless of their physical location. These platforms support the Agile principle of face-to-face interaction, albeit in a virtual format, fostering a sense of cohesion and shared purpose among team members.

Version control systems, such as Git and Subversion, have become critical for managing code repositories in Agile environments. These systems enable multiple developers to work on different features simultaneously while maintaining a single source of truth for the codebase. Branching and merging strategies supported by version control systems allow for parallel development and integration, aligning with the iterative nature of Scrum.

The integration of DevOps practices with Agile methodologies has further streamlined the software development lifecycle. DevOps emphasizes collaboration between development and operations teams, promoting a culture of shared responsibility for the entire product lifecycle. Tools like Docker and Kubernetes facilitate containerization and orchestration, ensuring that applications are scalable, portable, and resilient. This synergy between Agile and DevOps practices enhances the ability to deliver value continuously and reliably.

These technological advancements have significantly impacted the effectiveness and efficiency of Scrum and the Agile Planning Onion. By leveraging cutting-edge tools and practices, Agile teams can navigate the complexities of modern software development, delivering high-quality products that meet the evolving needs of stakeholders.

The Future of Remote Work

As organizations increasingly adopt Agile methodologies, the landscape of work continues to evolve, particularly with the rise of remote work. The Agile Planning Onion, a framework for iterative and incremental planning, provides a structured approach that can seamlessly integrate into remote work environments. The implications of this integration are profound, affecting team dynamics, productivity, and overall project success.

The transition to remote work necessitates a reevaluation of traditional Scrum practices. Daily stand-ups, sprint planning, and retrospectives must adapt to virtual formats. Tools such as video conferencing, instant messaging, and collaborative software become indispensable. These technologies facilitate communication and ensure that team members remain aligned with project goals. The Agile Planning Onion's layered approach to planning—ranging from strategic to daily tasks—can be effectively managed through these digital tools, promoting transparency and accountability.

One significant advantage of remote work within the Agile framework is the potential for increased flexibility. Team members can work from various locations and time zones, allowing for a more diverse and inclusive workforce. This geographical dispersion, however, introduces challenges in synchronizing efforts and maintaining a cohesive team culture. To mitigate these issues, it is crucial to establish clear communication protocols and regular check-ins. Virtual team-building activities can also foster a sense of camaraderie and trust among remote team members.

The productivity of remote Agile teams can be influenced by several factors, including the home working environment, access to resources, and individual work habits. The Agile Planning Onion's iterative nature allows for continuous assessment and adjustment of workflows to optimize productivity. Regular retrospectives provide opportunities to identify and address any impediments hindering remote work efficiency. By leveraging data and feedback, teams can implement targeted improvements, ensuring that productivity remains high.

Remote work also impacts the role of the Scrum Master and Product Owner. The Scrum Master must be adept at facilitating virtual meetings and ensuring that team members remain engaged and focused. They must also be vigilant in identifying and resolving any issues that may arise due to the remote setting, such as communication breakdowns or feelings of isolation among team members. The Product Owner, on the other hand, must ensure that the product backlog is meticulously maintained and accessible to all team members, regardless of their location. Clear and concise user stories, along with well-defined acceptance criteria, become even more critical in a remote context.

The future of remote work in Agile environments will likely see further advancements in technology and collaboration tools. Artificial intelligence and machine learning could play a role in automating routine tasks and providing predictive insights, thereby enhancing the planning and execution process. Virtual reality and augmented reality may offer new ways to conduct immersive meetings and collaborative sessions, bridging the gap between physical and virtual workspaces.

As organizations continue to navigate the complexities of remote work, the Agile Planning Onion offers a robust framework for maintaining structure and focus. By embracing the principles of Agile and leveraging the appropriate tools and techniques, remote teams can achieve high levels of performance and deliver exceptional results. The evolution of remote work will undoubtedly bring new challenges, but with a solid foundation in Agile practices, teams can adapt and thrive in this dynamic landscape.

Continuous Learning and Adaptation

The iterative nature of Scrum and Agile methodologies inherently supports an environment conducive to continuous learning and adaptation. This chapter delves into the mechanisms and practices that facilitate this ongoing process, ensuring that teams remain agile and responsive to evolving project requirements and environmental changes.

Central to continuous learning in Scrum is the feedback loop, which is embedded in the framework through regular ceremonies such as Sprint Reviews and Retrospectives. During Sprint Reviews, teams present their work to stakeholders, receiving critical feedback that informs subsequent iterations. This iterative feedback mechanism ensures that the product increment aligns with stakeholder expectations and market demands, thus fostering a culture of continuous improvement.

Retrospectives, on the other hand, serve as introspective sessions where team members reflect on the Sprint's outcomes and processes. By identifying what went well, what didn't, and areas for improvement, teams can adapt their practices and workflows. This practice not only enhances team efficiency but also promotes a culture of transparency and collective responsibility.

The role of the Scrum Master is pivotal in facilitating continuous learning and adaptation. By guiding the team through the Scrum process and ensuring adherence to Agile principles, the Scrum Master helps the team navigate obstacles and optimize their workflow. Moreover, the Scrum Master encourages a mindset of experimentation, where teams are empowered to try new approaches and learn from their outcomes. This experimental mindset is crucial for fostering innovation and resilience within the team.

In addition to the formal Scrum ceremonies, continuous learning is supported by various Agile practices such as pair programming, code reviews, and automated testing. Pair programming, where two developers work together at one workstation, promotes knowledge sharing and cross-training. Code reviews provide an opportunity for team members to learn from each other's expertise and ensure code quality. Automated testing, by providing immediate feedback on code changes, allows teams to quickly identify and rectify issues, thus maintaining a high standard of quality.

The Agile Planning Onion, as a conceptual model, further facilitates continuous learning and adaptation by providing a structured approach to planning at multiple levels. At the core of the onion is the daily planning, which involves the Daily Scrum meeting. This meeting allows teams to adapt their plan based on the previous day's progress and emerging insights. Moving outward, the Sprint Planning meeting sets the stage for short-term goals, while Release Planning and Product Roadmap provide a longer-term vision. Each layer of the onion represents an opportunity for feedback and adaptation, ensuring that plans remain relevant and actionable.

The incorporation of metrics and data-driven decision-making is another critical aspect of continuous learning in Scrum. By tracking key performance indicators (KPIs) such as velocity, cycle time, and defect rates, teams can gain insights into their performance and identify areas for improvement. These metrics enable teams to make informed decisions and adjust their strategies accordingly.

Furthermore, continuous learning extends beyond individual teams to the organizational level. Organizations adopting Scrum and Agile methodologies often establish Communities of Practice (CoPs) where practitioners can share knowledge, experiences, and best practices. These communities foster a culture of collective learning and continuous improvement across the organization.

Continuous learning and adaptation are fundamental to the success of Scrum and Agile methodologies. Through structured feedback mechanisms, supportive roles, collaborative practices, and data-driven insights, teams can continuously refine their processes and deliver high-quality products that meet stakeholder needs. The Agile Planning Onion provides a comprehensive framework for planning and adaptation at various levels, ensuring that teams remain responsive and resilient in a dynamic environment.

www.ingramcontent.com/pod-product-compliance
Lightning Source LLC
Chambersburg PA
CBHW071921210526
45479CB00002B/507